LOW IMPACT DEVELOPMENT AND SUSTAINABLE STORMWATER MANAGEMENT

Thomas H. Cahill, P.E.

Consultant
Water Resources Engineering

A JOHN WILEY & SONS, INC., PUBLICATION

Published by John Wiley & Sons, Inc., Hoboken, New Jersey
Published simultaneously in Canada

For general information on our other products and services or for technical support, please contact our Customer Care Department within the United States at 877-762-2974, outside the United States at 317-572-3993 or fax 317-572-4002.

Wiley also publishes its books in a variety of electronic formats. Some content that appears in print may not be available in electronic formats. For more information about Wiley products, visit our web site at www.wiley.com.

Library of Congress Cataloging-in-Publication Data:

Cahill, Thomas H., 1939–
Low impact development and sustainable stormwater management / Thomas H. Cahill.
p. cm.
Includes index.
ISBN 978-0-470-09675-8 (cloth)
1. Urban runoff—Management. 2. Sustainable urban development. I. Title.
TD657.C34 2012
628′.212—dc23

2011037607

10 9 8 7 6 5 4 3 2 1

CONTENTS

PROLOGUE: HABITAT, SUSTAINABILITY, AND STORMWATER MANAGEMENT

Over the past 4.5 million years, as our species has evolved from a simple mammal that learned to walk upright, we have sought suitable habitat. For most of that period, the form of this living space was quite simple, with only two basic criteria: protection from the weather, and a source of water. Within the past 12,000 years, our concept of what constitutes habitat has evolved and become infinitely more complex, although these basic criteria have remained unchanged. From caves or other natural shelters to structures built with local vegetation, the transformation to more elaborate buildings for habitat, commerce, worship, and recreation has taken place in a relatively brief period of our existence as sentient creatures.

As we increased in numbers and the fabric of social structure evolved, our perspective of the local environment did not change. The world and everything in it was a gift from our god (or gods), and natural resources were to be exploited as necessary to serve our needs. Even during the current period from the founding of the United States of America, the underlying dynamic of development was to conquer the wilderness, clear the forest, fell the ancient trees, dam the mighty rivers, and till every possible acre for food productivity, especially along river valleys rich with deposited sediment. The gifts of nature were abundant and available, and the land belonged to each property owner, to use as he or she saw fit.

For future generations, the beginning of the twenty-first century may well be considered the age of environmental enlightenment, when the extremes of energy and water exploitation that characterized the twentieth century have finally been recognized, and alternative strategies formulated. One of the most important concepts to have evolved in the past two decades is that of *sustainability*, which in the context of land development means the ability to construct our needed facilities without destroying the land and water systems that are essential elements of our habitat. We are only beginning to comprehend that if we do not sustain these natural resources for future generations, our communities will collapse within the near future. Countless examples of such failures can be drawn from previous societies, but nothing on the scale presently anticipated.

It serves no useful purpose to dwell on "doomsday" scenarios to illustrate this potential collapse. This book develops simple and practical examples of designs that change present practice without sacrificing any of the desired comforts of a built environment. It is essentially a positive response to the issues at hand, intended to influence the current generation of engineers, architects, landscape architects, planners, and developers who will build our future habitats. This process will follow new methods and use new materials to create structures that shelter us from the elements, assure a safe and sufficient supply of water, and provide opportunities for our children to do the same. This concept is advocated in the Living Building program developed by the Cascadia Division of the U.S. Green Building Council, and provides a template for the future of building, with zero net water (and energy) as the basic design goals.

THOMAS H. CAHILL

ACKNOWLEDGMENTS

This book is a compilation of information developed over a period of 49 years, especially the past 35, as a number of stormwater management concepts began to evolve into a body of practice. These concepts center on the reduction of runoff volume, rather than simply runoff detention as has been the general method of dealing with the impact of development since the 1970s. As site designs evolved, broader questions were raised, such as how and where we should (and should not) build our structures. In time, stormwater management became part of a larger effort to rethink how we develop the land. A mix of disciplines contributed to these concepts, including civil and water resource engineering, landscape architecture, planning, and architecture, and it is accurate to say that the process is still evolving.

A mix of talent, drawn largely from the staff of Cahill Associates (no longer in practice), contributed many of the ideas and designs included here. Most important is Michele Adams, P.E., my daughter and partner, whose creative thinking is blended throughout the book and reflected in the quality of the work; Wesley R. Horner, P. P., the principal author of Chapter 4; Andrew Potts, P. E., who wrote most of Chapter 6; Daniel Wible, P. E., who designed many of the example projects illustrated throughout the book; and Tavis Dockwiler, L. A., principal of Veridian Design, who has been my muse in the somewhat confusing world of vegetation, following the role played initially by Carol Franklin, L. A., of Andropogon Associates. Other contributors include Richard Watson, P. P., who crafted much of Chapter 5, and Charles Miller, P. E., president of Roofmeadows, Inc., whose pioneering work in bringing the practice and construction of vegetated roofs to the United States serves as an example of how good ideas succeed if you are tenacious.

Cahill Associates, Inc. (CA) was acquired by CH2M HILL (CH2) in 2008, and the same team continued to work on then-contracted and new projects. The case studies described in Appendix B were performed primarily under the guidance of the CA firm, but one new study, the Allegheny Riverfront Vision Plan, was carried out by CH2 under the project management of Courtney Marm, P. P.

A special thank you goes to my daughter Christine Steininger, who played a critical role in the final production of graphics for this book, with a skill level far beyond my own.

T. H. C.

1

RAINWATER AS THE RESOURCE

1.1 THE WATER BALANCE AS A GUIDE FOR SUSTAINABLE DESIGN

In every portion of the planet, the cycle of water provides the same natural model: The water resource is replenished with each season and the land surface responds to this cycle of abundance or drought with a vegetative system that flourishes and diminishes with the available rainfall. The hydrologic cycle is continuous, but it is by no means constant, and every human habitat must recognize and live within the limits and constraints of this dynamic process. Over the past 4.5 million years, our species has learned to live in balance with the water cycle; or if it changes over time, migrate to other environments.

Unfortunately, over the past century, our modern society has not followed this process in the building of our current communities. As our numbers increased and spread across the land surface, we began to exploit rather than sustain our land and water resources. During the past century, our control of energy sources allowed us to neglect the principle of sustaining our habitat, and we gave little thought as to how we built our modern cities, disregarding the local environment and the natural limits of each place. Guided by a false confidence that we could conquer any constraint or natural limitation, we have stripped and sculpted the land to fit our perceived image of how we can best situate our structures. We have exploited the available water resources, without careful consideration of where we live in terms of natural topography and hydrology.

Low Impact Development and Sustainable Stormwater Management, First Edition. Thomas H. Cahill.

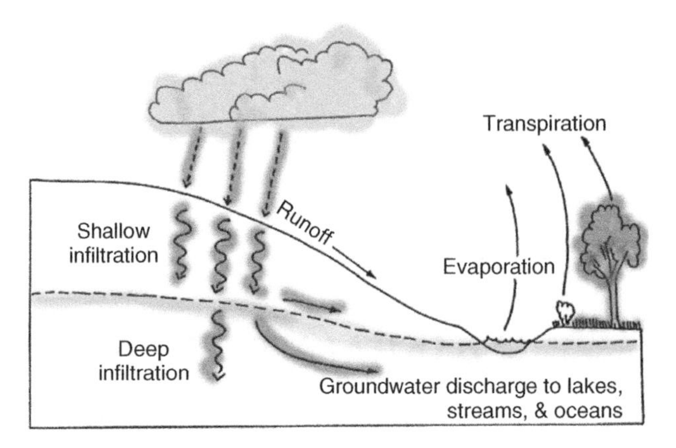

Figure 1-1 The hydrologic cycle.

The *hydrologic cycle* or *water balance* serves as a model for understanding the concept of sustainability of our water resources (Figure 1-1). The challenge of sustainability is to draw upon elements of this cycle to serve our needs without significantly disrupting the balance. With careful land use planning and water resource management, every available drop of rain can be used and reused to meet our needs without destroying the quality or affecting the character of natural streams and rivers. Many of our uses, such as drinking supply, can be largely recycled with the proper waste system design, and many other uses can be reduced in quantity if they are largely "consumptive" uses, such as irrigation of artificial landscapes. Consumptive demands of cultivation can also be reduced by methods such as drip irrigation, and energy systems can be designed that do not consume fresh water in the cooling process. All modern water supplies require energy, and most energy systems affect water. Similar to the land–water dynamic, the energy–water interrelationship requires that any system changes consider both resources.

The principle of water balance is best understood in the context of a measurable land area—watershed, drainage basin, or land parcel—that quantifies the water cycle. The rain that falls on the land surface over a period of time defines the magnitude of the resource and the quantity required to sustain the cycle. The potential demands on this balance imposed by our land development process can then be applied to this model as an initial step in understanding how the cycle should guide our activity on the land.

Perhaps the easiest way to understand the concept of the water balance is to consider a small unit area (Figure 1-2), an acre or hectare, and measure the movement of rainfall through this tiny portion of the planet. The flow begins (or continues) with rainfall, shown in the figure as the annual average for a temperate climate, the mid-Atlantic region of eastern North America. Whereas the annual amount of rainfall varies greatly from place to place across the United States (Table 1-1) and can also experience significant seasonal differences (Table 1-2), the hydrologic cycle remains a constant.

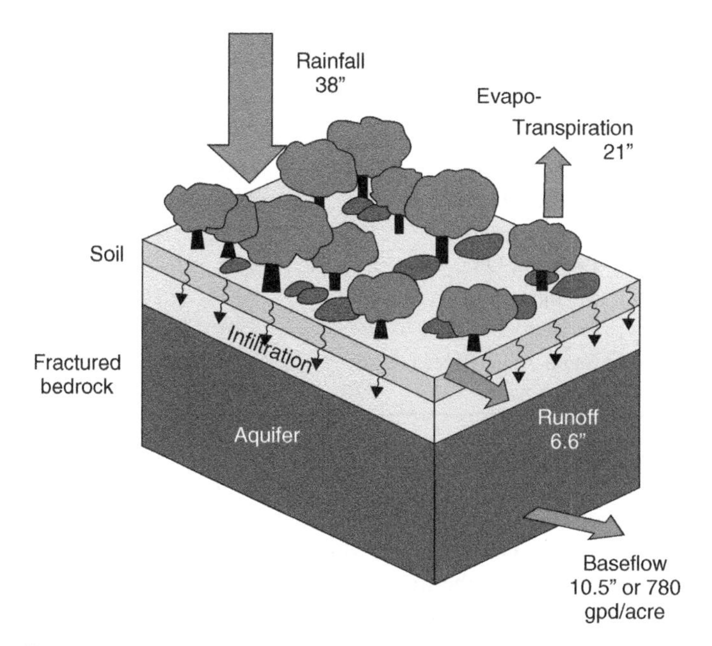

Figure 1-2 The hydrologic cycle on an undeveloped unit area (in./yr).

Table 1-1 Annual Rainfall in Major U.S. Cities

City	Annual Rainfall (in.)	Annual Snowfall (in.)
Albany, NY	38.6	64.4
Anchorage, AL	16.1	70.8
Atlanta, GA	50.2	2.1
Austin, TX	33.6	0.9
Boston, MA	42.5	42.8
Charlotte, NC	43.5	5.6
Chicago, IL	36.3	38.0
Denver, CO	15.8	60.3
Duluth, MN	31.0	80.6
Honolulu, HI	18.3	0
Houston, TX	47.8	0.4

Source: [1].

It is also possible to structure a more complicated model of this dynamic process (Figure 1-3), realizing that the water movement through all elements is continuous, while some elements, such as the soil mantle, act as short-term storage units, holding or releasing moisture from year to year. Other processes, such as evapotranspiration, vary greatly from season to season, and by location, throughout the year. Thus even this more complex graphic fails to fully describe the water balance cycle.

Table 1-2 Seasonal Variation of Rainfall in Regional Watersheds

Region	Major River Basin and City	Rainfall (in.)				
		Spring	Summer	Fall	Winter	Total
Northeast	Delaware River	11	11.3	10.6	9.7	42.5
	Philadelphia, PA	*26%*	*26%*	*25%*	*23%*	
Northwest	Willamette River	13.8	3.5	15.7	27.2	60.2
	Portland, OR	*23%*	*6%*	*26%*	*45%*	
Southeast	St. Johns River	10	10	19.1	12.2	51.2
	Jacksonville, FL	*19%*	*19%*	*37%*	*25%*	
Southwest	Santa Anna River	4	0.4	2.7	7.5	14.6
	Los Angeles, CA	*27%*	*4%*	*15%*	*51%*	

The development of plants on the planet surface long preceded the mammals from which we evolved, and plants have fulfilled their part in the hydrologic cycle for several billion years. On those land surfaces that evolve a natural vegetative cover, especially woodlands, the trees and grasslands utilize the input of rainfall to live by *photosynthesis* [2], drawing the infiltrating moisture from the soil (or directly from the atmosphere) and transforming the water into oxygen and organic matter, a process described by the reaction

$$(6H_2O + 6CO_2) + (\text{sunlight, 48 mol}) = 6O_2 + C_6H_{12}O_6$$

This simple miracle of plant life is carried out by the role of chlorophyll in the vegetation. In addition to producing the oxygen by which all species live, this process maintains the critical balance of CO_2 in the atmosphere for the benefit of all animal life forms, including the human species. While the air we breathe is comprised primarily of 78% nitrogen with slightly less than 21% oxygen, the role of minor gases (argon, 0.93%; CO_2, 0.038%) and water (1%) is critical in maintaining the temperature at a relatively constant level over time. The rapid increase in CO_2 over the past century has played an important and causal role in global warming, specifically as the result of burning fossil fuels [3, 4]. Thus, the importance of sustaining surface vegetation, especially trees, during the land development process cannot be overstated (Figure 1-4), as it compensates for this human impact [5, 6]. It should be noted that terrestrial vegetation provides only a portion of the photosynthetic production, with marine plankton actually generating more of the balance on a global basis.

On a naturally vegetated land surface, about half of the rain that falls is returned to the atmosphere by the evapotranspiration process. The balance of infiltrating rainfall, not utilized by the vegetation or evaporated from the surface by sunlight and air currents, infiltrates or percolates slowly (or quickly) into the soil mantle. A portion of this rain drains deep into the soil and weathered rock surface, eventually reaching the zone of saturation, described as the water table (Figure 1-5), and becomes groundwater.

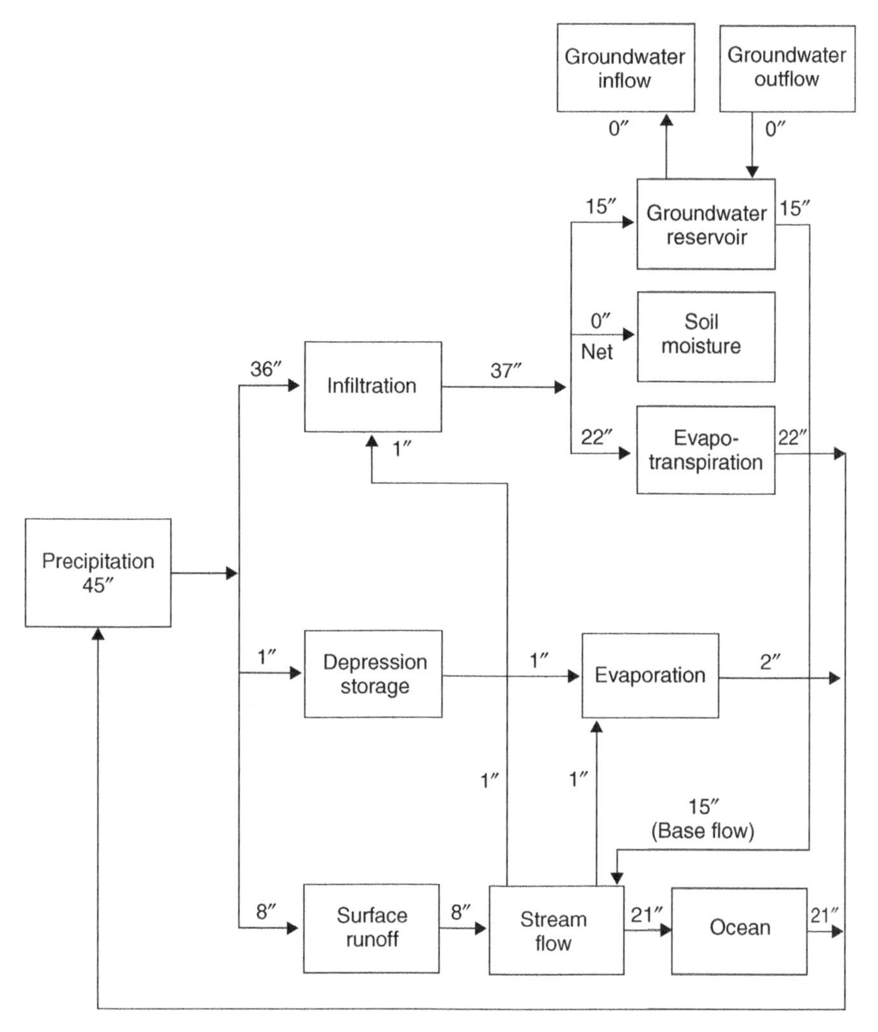

Figure 1-3 The hydrologic cycle or water balance model for a watershed in southeastern Pennsylvania: the Brandywine model project, 1984.

As each raindrop is added to this groundwater, it begins to move in the direction of available energy, created by the inexorable pull of gravity. Since the easiest pathway for displacement is through the soil (and fractures in the rock) following the surface of the land, this water eventually travels downhill, emerging as a seep or spring, flowing over the surface to a swale or steam channel. Actually, as each raindrop enters the groundwater, it displaces water from the low end of the saturated zone. A single raindrop may actually take weeks or months to complete the journey from where it falls on the land surface to the point of discharge downgradient, as it returns to the surface.

Figure 1-4 The perfect LID measure for stormwater management: a tree.

Of course, not all infiltrating rainfall follows an identical pathway of movement in the subsurface, and the complex layering of the soil in different horizons, each with a very different permeability, can make this journey lengthy and circuitous. Where highly impervious layers exist in the soil mantle, infiltrating rain will move across this surface, again following the energy gradient. The underlying bedrock also influences the speed and direction of groundwater movement, in both the unsaturated zone and deep below the water table. If the underlying rock is comprised of soluble carbonates, it includes open solution channels or subsurface flow pathways that formed hundreds of millions of years ago and now provide an underground river network, carrying the rainfall many miles from the

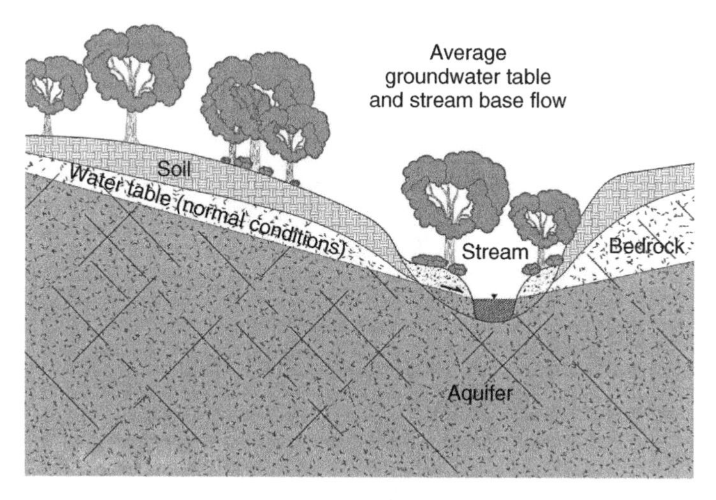

Figure 1-5 Groundwater recharge feeds the local surface waters and sustains base flow.

initial point of infiltration. In coastal watersheds, the groundwater may discharge directly to estuary systems, never reappearing on the land surface.

In some physiographic regions, a fraction of the infiltrating rain enters into deeper aquifers and does not reappear at the surface, but may remain stored for centuries. In active seismic regions, geothermal sources may actually bring some of the deep water to the surface. This vertical flow of groundwater may comprise a portion of surface systems, such as the Snake River tributary in the Columbia River system, originating from the "hot spot" that forms the geysers of Yellowstone (Figure 1-6). However, for most of the developed regions of the United States, the simple model illustrated by Figure 1-2 is a valid representation of this complex water balance.

While the full hydrologic cycle includes water movement on a global basis, the consideration of stormwater management is limited to the freshwater portion of the total resource, a fraction of the world's water (about 2.5%). Most of that fresh water is currently contained in ice (although the future is quite uncertain) and represents 77.2%, with an additional 22.6% contained in the subsurface as groundwater, leaving only 0.32% in surface rivers and lakes, 0.18% as soil moisture, and 0.04% in the atmosphere, for a total "available" water resource of 0.54% [8]. All of the following discussion is concerned with this sum of these small portions, although it amounts to the trillions of gallons of water that sustain the human biosphere.

1.2 THE WATER BALANCE BY REGION

Although the most obvious measure of the water resources available in a given region of the country is the average annual rainfall received, this statistic can be deceiving if we do not recognize the potential variability in this measure,

Figure 1-6 Deep groundwater is discharged by geothermal vents (Yellowstone). (From [7].)

especially in arid regions such as the Southwest, where the extremes of "wet" and "dry" years can result in a system that experiences a crisis under both cycles. It is the extremes of the cycle that create the greatest stress in every community, and the duration of individual droughts or flood-creating rainfall periods that measure how well or poorly we have built our communities.

Most large river basins in the United States have experienced significant human alteration or structural intervention over the past two centuries. It is interesting to consider the net effect of human activities on the regional watershed, although we have no baseline (pre-disturbance) flow data to compare with current conditions. However, we can compare the net runoff generated in these large systems with the rainfall experienced within the watershed (Table 1-3). Also shown is a reference city, usually situated at the downstream reach of the river basin. In the Santa Anna basin draining to Los Angeles, the inflow from three diversion canals affects these statistics significantly.

Table 1-3 Water Balance by Region and River Basin

Region and Percent of Area	River Basin and Reference City	Basin Size (m^2)	Mean Discharge (ft^3/sec)	Alteration	Rainfall (in)		Runoff (in.)
					City	Basin	
NE 60%	Delaware Philadelphia, PA	12,757	14,902	NYC diversion	42	42.5	25.6
NW 63%	Willamette Portland, OR	11,478	32,384	Dams/lakes	37	60.2	38.2
SE 23%	St. Johns Jacksonville, FL	8,702	7,840	Lakes	52	51.6	11.8
SW 2.3%	Santa Anna Los Angeles, CA	2,438	60	Diversions	15	13.4	0.3

Source: Derived from [9].

Whatever the average annual rainfall or variability of this volume in a given location, the design of structures or systems to convey or mitigate the impacts of this volume (and flow rate) of surface runoff have always focused on individual storm events. These "design storms" are events during which the intensity, duration, and amount of rainfall produce the most severe impacts.

We remember the most extreme rainfall events, especially when they are the result of cyclonic storm patterns produced in both the Atlantic and Pacific oceans that approach the mainland in the form of hurricanes or cyclones. We even identify them by name when they reach a given magnitude or anticipated wind speed, assigning a category of intensity that can change during the approach. Most recent memory cannot help to identify hurricane Katrina (Figure 1-7), which devastated the Gulf coast in September 2005, but other names and memories are shared by communities throughout the country. Most periods of prolonged rainfall do not receive this recognition or nomenclature, but have produced dramatic flooding impacts in large and small watersheds.

The statistic of rainfall that has the most common usage in defining severe rainfall events is the 100-year storm, which is the rainfall that occurs during a 24-hour period with a frequency of once in 100 years. This figure cannot, however, convey the full impact on a local watershed of more severe and intense rainfalls. For example, in July 2004 the Rancocas Creek in southern New Jersey was visited by a rainfall pattern [10] that dumped some 13 in. in some portions of this small (250 m^2) watershed (which has a 100-year rainfall frequency of 7.2 in.), in a pattern that was anything but uniform. The net result was the destruction of some 22 small earthen dams, built for various purposes, and significant property damage (but no loss of life).

This type of localized event can be visited on any portion of the country, regardless of our statistics and classification of storms, and is repeated all too frequently all across the globe. While the total rainfall is a given period and

Figure 1-7 Hurricane Katrina strikes the U.S. Gulf coast.

the intensity of that precipitation have much to do with the resulting impact, the hydrologic response of any given watershed is also a function of land cover conditions, especially vegetation, and season, with frozen ground producing some of the most severe runoff conditions during early spring in mountainous regions.

If we were to measure all of the rainfalls at a given location over a century, we would find that the vast majority were of very small magnitude (Figure 1-8). The pie chart in the figure shows rainfall distribution for southeastern Pennsylvania, with a total annual rainfall of 44 in./yr. The relative distribution is the same for most other regions, with most of the storms less than 3 in. in total rainfall, and offers insight as to the defining statistic for a stormwater volume reduction management strategy.

While the traditional focus of concern has been the extremes of rainfall or drought, the major portion of our precipitation actually occurs in smaller, more frequent events. In fact, in almost every major physiographic or climatologic region, the 2-year-frequency rainfall serves as the defining statistic for the stormwater management designs that are outlined in this book. This rainfall and

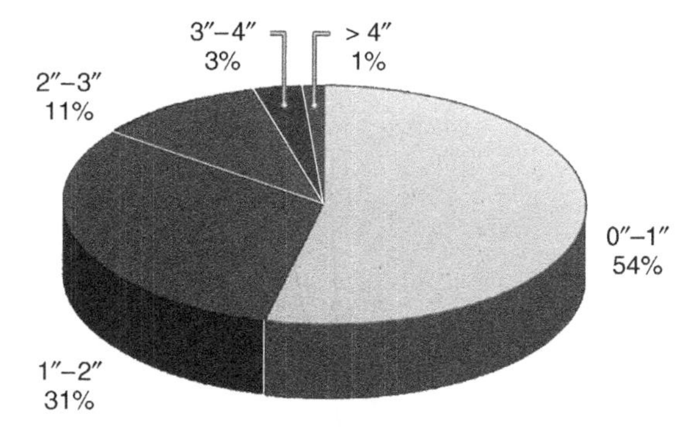

Figure 1-8 Frequency and magnitude of rainfall events, southeastern Pennsylvania. Most rainfall occurs in small storms, less than the 2-year frequency.

Table 1-4 Two-Year-Frequency Rainfall Event

U.S. Region	City	Two-Year Rainfall (in. in 24 hr)
Northeast	West Chester, PA	3.3
Northwest	Seattle, WA	3.2
Southeast	Chapel Hill, NC	3.6
Southwest	Los Angeles, CA	2.9
Central	Minneapolis, MN	2.5

that of all the storms of lesser magnitude represents about 95% of the total rainfall volume over a prolonged period of decades, and so better defines the efficiency of any proposed mitigation measure. Since this statistic has great significance as a basis for the design of most of the measures described in this book, it is important to compare the variation in this type of rain event in different portions of the country. Table 1-4 shows the 2-year-frequency rainfall in major regions, and the values are quite similar. Figure 1-9 illustrates the intensity of this type of storm over a 24-hour period (as well as the 100-year rainfall) for a mid-Atlantic watershed. This is described by an S-curve, developed by the Soil Conservation Service of the U.S. Department of Agriculture during the 1960s [11]. Of course, nature never cooperates with our assumptions concerning climate conditions such as rainfall patterns, but this type of distribution is assumed because it will produce the most extreme runoff conditions.

1.3 ARID ENVIRONMENTS: THE SOUTHERN CALIFORNIA MODEL

To be sustainable, low-impact design (LID) must consider all human demands on the hydrologic cycle that result from the land development process. This means that we begin our site planning with the issue of water supply, the single

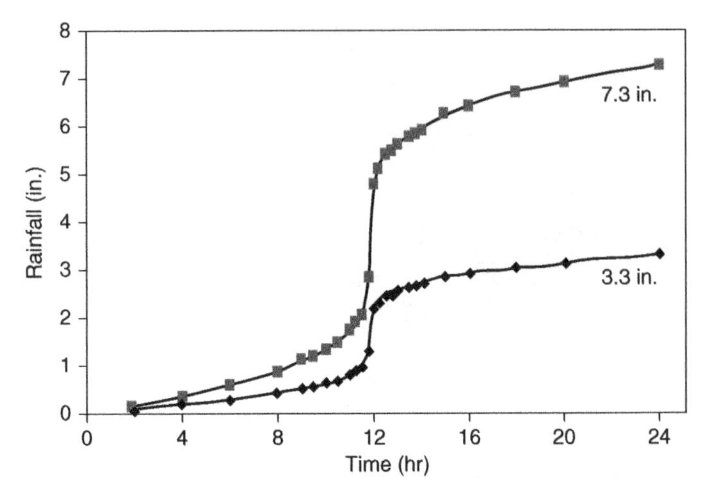

Figure 1-9 The S curves of assumed rainfall intensity and distribution; 2- and 100-year-frequency storms in southeastern Pennsylvania.

most critical aspect of site development. When this basic need is satisfied, we consider the return of this water to the cycle, containing all of the pollutants we have added during our use. The need for increasingly efficient pollutant removal processes during the past century has resulted largely from the increase in population and density in our land development, as we realized that the sewage from one community is the water supply for downstream residents.

Stormwater has been regarded as a nuisance, to be drained away from our developments as quickly as possible following a rainfall. In arid environments, the value of rainfall to support our continued occupation of habitat is an unutilized resource, especially when the available water supply is limited and runoff is discharged to coastal waters. Nowhere is this lack of water resource management more apparent than in southern California. In these coastal watersheds, land developments draining to the Pacific coast and to inland waters and reservoirs have generated significant increases in stormwater runoff volume, which in turn has contributed to the discharge of pollutants into receiving waters, degraded aquatic habitat, affected the recreational use of these waters, and interfered with their use as water supply. Through implementation of LID practices, these pollutant discharges can be reduced significantly, so that the quality of these coastal and inland waters can be restored and sustained.

But the potential of LID goes well beyond reducing the volume of polluted stormwater runoff. This rainfall can also be understood as a lost resource in the semiarid environment of southern California, where increasing demand for fresh water requires costly importation of water supplies to sustain ever-growing communities. Interestingly, the quantity of water imported into southern California is almost equal to the net loss of stormwater runoff to coastal waters (Table 1-5), referred to as the *salt sink*.

Table 1-5 Water Balance in Southern California, 2000 (1,000 Acre-Feet)

Supply		Use (Demand)	
Precipitation	7,500	Evapotranspiration	7,441
Imported	2,991	Consumptive use	1,819
Depletion of groundwater	1,245	Outflow to the salt sink	2,498
Total supply	**11,752**	**Total use**	**11,758**

Note: The runoff lost to the ocean is almost equal to the required import.

Could LID practices such as capture and reuse of stormwater runoff significantly reduce the need for importation of water supply to the region? These practices offer the possibility of working to redress at least some of the water cycle imbalances that confront southern California communities, but will require a significant rethinking of existing stormwater and water supply system designs.

Central to the concept of watershed sustainability is the water cycle and its balance, very roughly defined as the matching of water inputs ("supply") to water outputs ("demands"). Any watershed, physiographic region, or land area that can be well defined in terms of water supply and use, both natural and human, can also be evaluated in terms of water cycle. Analysis of the existing water balance for southern California demonstrates that the natural water resources are insufficient to meet the demands of the existing 19 million residents in the 11,000-square mile region from Ventura to San Diego, with 5 million additional residents projected to arrive by 2020 [12].

The deficit in natural water resources has been met over the years by three aqueducts (Figure 1-10), which convey imported water hundreds of miles to the region. Table 1-5 provided a simplification of this water resource balance for the year 2000, with both supply and demand (use) summarized from more detailed statistics. In 2000, rainfall provided only 72% of the total water supply use or demand, again with the quantity of runoff to the ocean [2,498 thousand acre-feet (TAF)] close to the water importation (2,991 TAF).

The depletion of groundwater is especially troublesome in this "balance." Although a number of large aquifer systems lie beneath the surface of the region and are constantly being replenished by recharge from surface sources, their capacity has been exceeded during most years. The Los Angeles area receives over 40% of their current water supply from these aquifers, utilizing spreading grounds to recharge both runoff and recycled effluent, based on the concept of "conjunctive use." However, year after year, the groundwater reservoirs are further depleted, despite our plans.

The Energy Demand for Water in Southern California

It requires over 10,200 kilowatthours (kWh) for every million gallons (MG) of water imported into southern California, 40 times greater than the national average and 20% of total residential energy usage for the region, as shown in

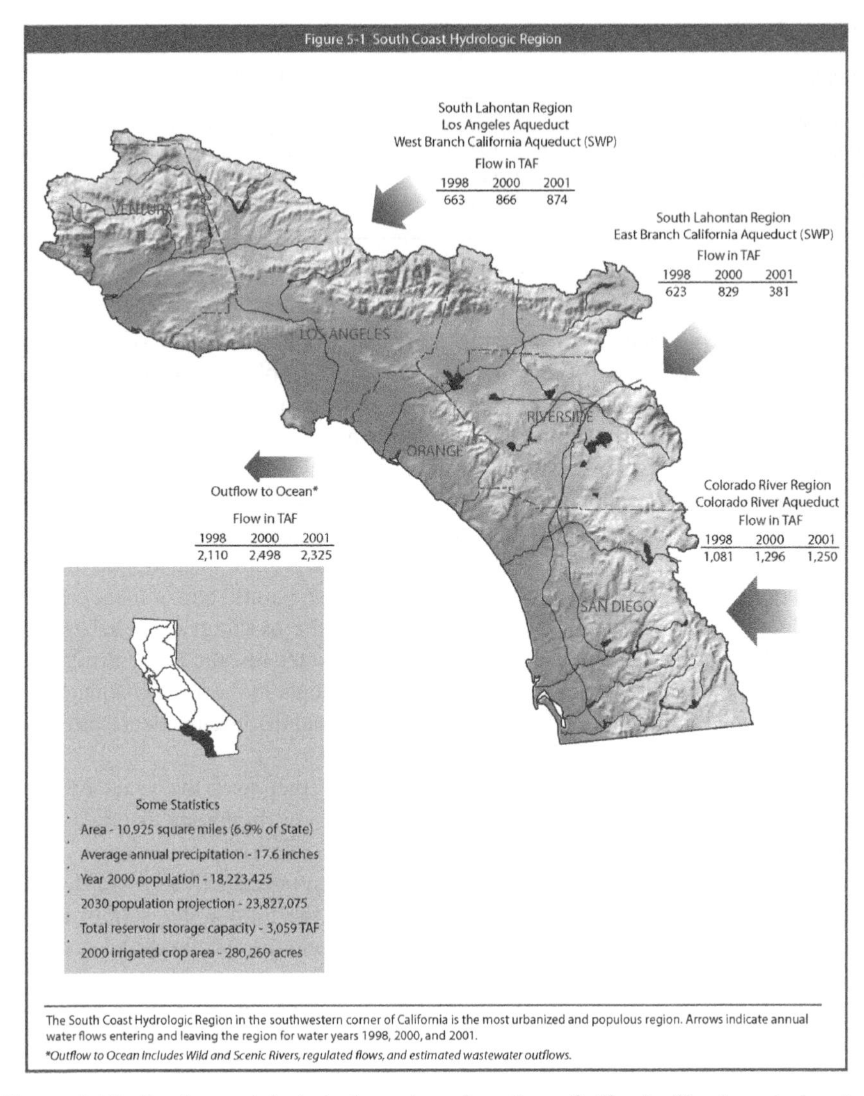

Figure 1-10 South coast hydrologic region of southern California. The imported water equals the runoff lost to the ocean.

Table 1-6. This can be compared with the national average of 250 kWh/MG. Any significant reduction in importation of water to the region can be expected to translate into significant energy savings.

What does energy demand have to do with LID? When we consider the application of sustainability concepts to land development, we must include the impact on both water and energy, and in this region the shortage of both resources is critical. If we can capture the rainfall and at the same time reduce the amount of

Table 1-6 Energy Demand of Water Supply in Southern California

Supply and conveyance	8,900 kWh/MG
Treatment and disposal	1,300 kWh/MG
Total	10,200 kWh/MG
National average	250 kWh/MG

Source: [13].

energy required to deliver imported water supplies, we will address both issues with one set of design solutions.

In southern California, this water cycle varies considerably. For example, in terms of annual precipitation, the total amount of rain received by the major communities in southern California varies dramatically; relatively wet Pasadena in the northern portion of the region receives twice the rainfall of semiarid San Diego in the southwest (20 in Pasadena, 16 in./yr in Riverside, 15 in./yr in Los Angeles, and 10 in./yr in San Diego). This variability is reflected in the size of storm events, which is so important in stormwater management calculations and therefore important to LID design. For example, the 2-year-frequency storm varies from 1.6 in. of rain in 24 hours in San Diego to 3.5 in. in Pasadena. This frequency rainfall comprises some 93 to 95% of all the rain that occurs in a century, so it is a defining statistic. Within that rainfall pattern is an equally varied difference, on both an annual and seasonal basis, with a well-defined "wet" and "dry" season. Such natural variability in the hydrologic cycle within the region makes stormwater management that much more challenging.

Of course, the fact that water is limited in southern California does not preclude the occurrence of coastal storms from the Pacific causing rainfall events that drop almost the total annual average rainfall over a period of a few days, as occurred during mid-January 2010. From the 18th through the 22nd, a total of 8 in. of rain fell in the Los Angeles region, causing widespread flooding, erosion and mud slides, evacuations, and most significantly, the structural failure of a number of residences perched precariously on hillsides. The local building codes allow the placement of buildings in locations that would not be considered sustainable in a less arid environment, where saturated soils are an assumed design condition. This cycle of rainfall, runoff, and mud slides was repeated in January 2011, and promises to occur frequently until land development planning better manages rainfall.

The current hydrologic cycle in southern California bears little resemblance to the natural system of a century ago, as so well described by D. Green in her book *Managing Water: Avoiding Crisis in California* [14]:

> In the 1920's, roughly 95% of the rainfall falling on Los Angeles either infiltrated into the ground or evaporated. Only 5% ran off to the sea. Today, with the extensive development and the paving over of our urban environment (as much as 80% of the land is now covered with roofs, roads, parking lots, patios, etc.) and the construction of the massive storm drain channel system, about 50% of stormwater runs off in the Los Angeles River drainage area, while 50% either infiltrates or evaporates. . . .

> About 90% of the San Gabriel River's flow is captured for recharge into the groundwater supply. On average, only about 20% of the upper Los Angeles River native runoff is captured, due to a lack of sufficient spreading capacity and the prevalence of clay soils.

The state Water Board has developed models of the current reality of water balance in southern California, and these are a far cry from simplistic illustrations. Obviously, no single model adequately describes the existing cycle of water in all of the region's watersheds and the importance of stormwater in the cycle. Based on the current reality, LID must be integrated in the land development and redevelopment process to use the limited rainfall received most efficiently to sustain the daily water demand of 140 gallons per person, even as imported supplies diminish. At the same time, we must restore those elements of the natural system that remain or can be sustained. Even this statistic of per capita consumption must be reduced, and that reduction turns on ways to sustain our landscapes with less water, using stored rainfall until we can change the landscape paradigm to a more natural vegetative palette over the next decade.

1.4 THE ALTERED WATER BALANCE AND HYDROLOGIC IMPACTS

Imperviousness

How do our current development practices affect the hydrologic cycle? The most obvious alteration is the use of impervious materials to build our communities, which for virtually all of the past 10 millennia consisted only of the rooftops of our buildings, temples, and homes. Until the beginning of the twentieth century, the land surface surrounding building structures was always comprised of compacted soil or paving stones in one form or another (Figure 1-11), with only a few major cities utilizing pavement surfaces that could be considered impervious.

This building pattern was totally altered a century ago, when the battle was concluded over which energy system would be used to power the new horseless buggy. The internal combustion engine won out over the steam engine, and gasoline became the fuel to power the rapidly expanding demand for transportation vehicles. This production in the "cracking tower" of petroleum refining resulted in an accumulation of "tar" in the bottom of the tower, similar in property to natural occurring tar pits mined in a few locations around the world. It was soon discovered that this residue of gasoline manufacturing could be combined with the graduated stone roads of the previous century (macadam) to form Tarmac surfaces, well suited for paving of our urban streets and roadways (Figure 1-12). As the demand for gasoline skyrocketed, the availability of the newly defined "asphalt concrete" made the surfacing of roads affordable, and so began a cycle of increased imperviousness that characterized the land development process throughout the twentieth century.

Figure 1-11 The urban environment in the eighteenth century (1799). The road was pervious.

A popular tune of the 1970s declared our wish to "... *pave paradise and put up a parking lot* ..."—not an exaggeration but an accurate description of this transformation of the land surface (Figure 1-13). Recent studies of land development patterns in developing watersheds have distinguished the different types of impervious cover that form the current patterns of land cover, and these land cover analyses generally conclude that about 25% of the impervious cover found in a watershed is comprised of the building structures serving human activities, and 75% is comprised of impervious pavements, largely asphalt concrete, but also portland cement concrete, that serve for the transport and storage of our automotive vehicles. Other studies [15] have concluded that when a watershed approaches an impervious cover of 25%, the water resources have been so altered that water quality and quantity have been severely degraded.

Countless small stream systems, especially in the headwaters of larger river basins, have been developed, filled, built upon, piped, and paved over during the past century. A good example is a small (440-acre) stream called Meeting of the Waters Creek in Chapel Hill, North Carolina, a tributary of the much larger Flint River basin, which drains to the coastal Atlantic estuary. Present conditions reflect expansion of the University of North Carolina over the past five decades. The present land cover is 77% impervious and still increasing (Figure 1-14A), with

Figure 1-12 The urban environment 200 years later (1990), with impervious pavements.

Figure 1-13 A typical commercial site, with the building surrounded by a sea of impervious pavement.

(A)

(B)

Figure 1-14 (A) The University of North Carolina campus in Chapel Hill was over 70% impervious by 2001. (B) Meeting of the Waters Creek underlies the campus and emerges from sewers below.

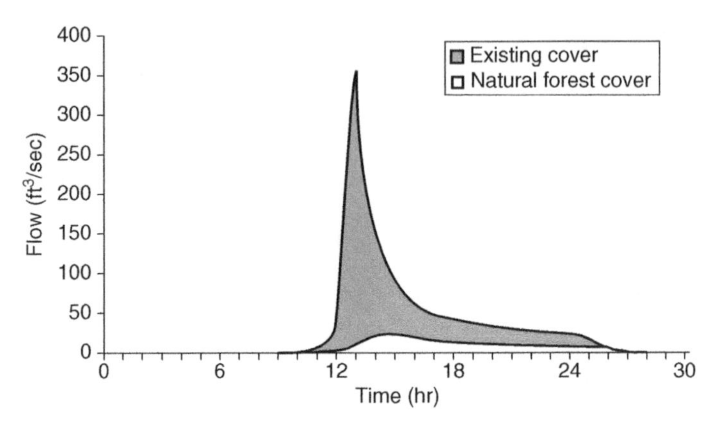

Figure 1-15 Hydrograph of Meeting of the Waters Creek, as affected by impervious surfaces.

the original stream system replaced by a network of sewer pipes that discharge to the remnant stream channel (Figure 1-14B).

The increase in runoff volume produced by this increase in impervious cover is dramatic (Figure 1-15), as illustrated during the 2-year-frequency rainfall. The increase in runoff volume (the area under the hydrograph) affects the downstream riparian corridor, eroding stream banks and conveying the pollutant load from the impervious surfaces as well as the channel to the large Lake Jordan, a future water supply reservoir, several miles downstream [16].

Impervious surfaces, be they rooftops, pavements, or streets, turn every drop of rainfall into direct and immediate runoff. In addition, the recharge of groundwater aquifers is effectively reduced or prevented altogether. Finally, the energy of runoff scours every pollutant that is dripped, dropped, spilled, or spread on the impervious (and pervious) land surfaces, and convey this **non-point source** pollution to surface waters. In all regions of the country, this impact can be understood by comparing the runoff volume increase from pre- to postdevelopment. For example, in the mid-Atlantic, where the annual rainfall reaches 45 in. and natural runoff is on the order of 8 inches, the net increase in runoff from impervious surfaces is 36 in. (3 ft) per year. One need only consider a typical parking lot surface as shown in Figure 1-13, and picture it covered with this depth of increased runoff, to fully understand the underlying cause of stormwater impacts from land development.

Increased Volume of Runoff

Figure 1-16 represents the same unit area as that shown in Figure 1-2, and illustrates the net impact of land development and impervious surfaces on the hydrologic cycle. To describe the process in simple terms, any impervious surface results in direct rainfall being converted into immediate and almost total runoff. Our building programs have always tried to transport this runoff away from the

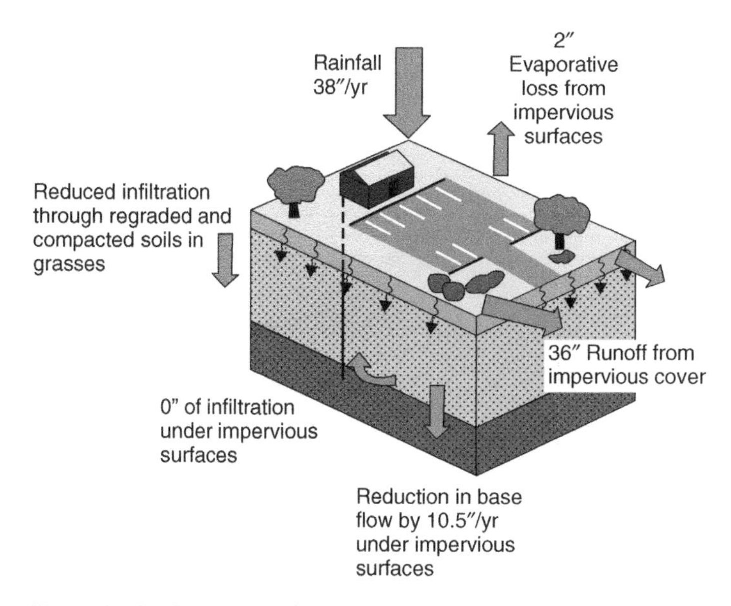

Figure 1-16 The impact of land development on the hydrologic cycle.

built environment to the remnant surface flow pathways and streams as quickly and directly as possible and to assure that both buildings and roadways can be utilized in the most severe rainfalls. Although this makes a great deal of sense from the building and transportation perspective, the net result was and is a dramatic and rapid increase in the rate and volume of runoff from a developed landscape. Observing this effect from a position downstream of the developed land, Figure 1-17 illustrates the rise in stream flow produced by a sudden increase in runoff from the upland drainage. This *hydrograph* of flow rate versus time defines a volume of runoff produced within a river basin, watershed, or catchment as the area contained within the figure, and is the most important aspect of the hydrograph measure.

Regardless of where land development occurs, the increases in imperviousness, the changes in vegetation, and the soil compaction associated with that development result in significant increases in runoff volume. The relative increase in runoff volume varies with event magnitude (return period). For example, the 2-year rainfall of 3.27 in. each 24 hours in southeastern Pennsylvania will result in an increase in runoff volume of 2.6 in. from every square foot of impervious surface placed on well-drained HSG B soil in woodland cover (Figure 1-18). For larger events, as the total rainfall increases, the net runoff also increases, but less than proportionately. For example, total rainfall for the 100-year storm is twice the rainfall for the 2-year storm (7.5 in. vs. 3.27 in.); however, the increase in runoff is only 1.7 in. (4.3 – 2.6 in.). This pattern holds true throughout the United States.

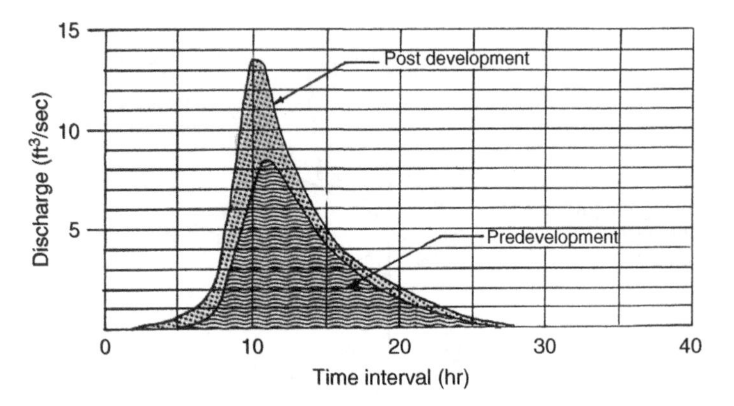

Figure 1-17 Hydrograph of increase downstream of development.

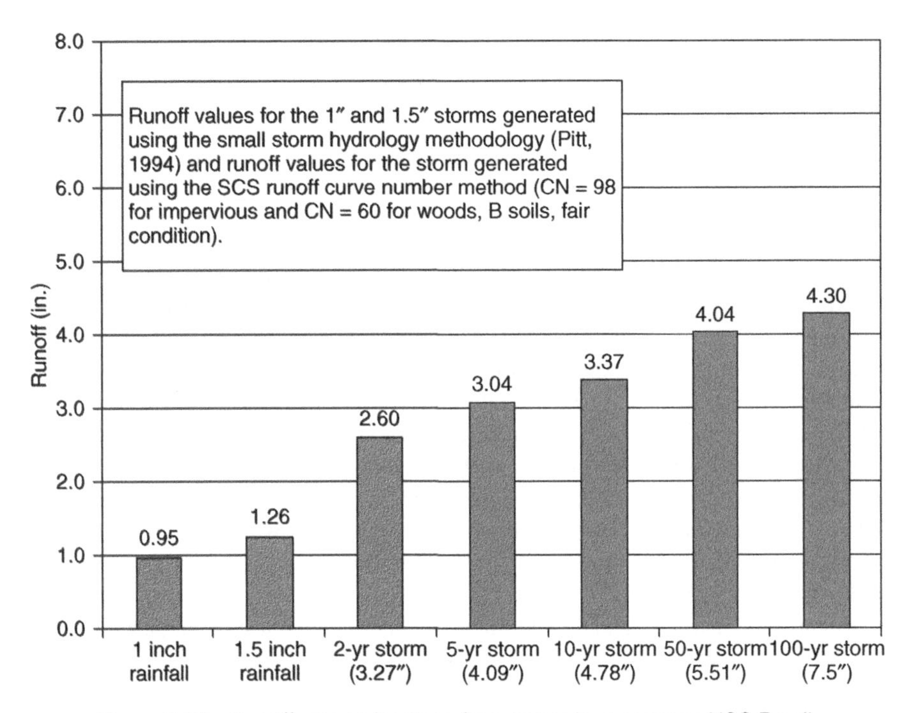

Figure 1-18 Runoff volume increase from impervious cover on HSG B soil.

For a specific site, the net increase in runoff volume during a given storm depends on both the predevelopment permeability of the natural soil and the vegetative cover. Poorly drained soils result in a smaller increase of runoff volume because the volume of predevelopment runoff is already high. Therefore, the amount of runoff resulting from development does not represent as large a net increase. Using the same rainfall values, Figure 1-19 illustrates that the 2-year rainfall of 3.27 in. in 24 hours produces an increase of only 2.01 in. on a HSG C

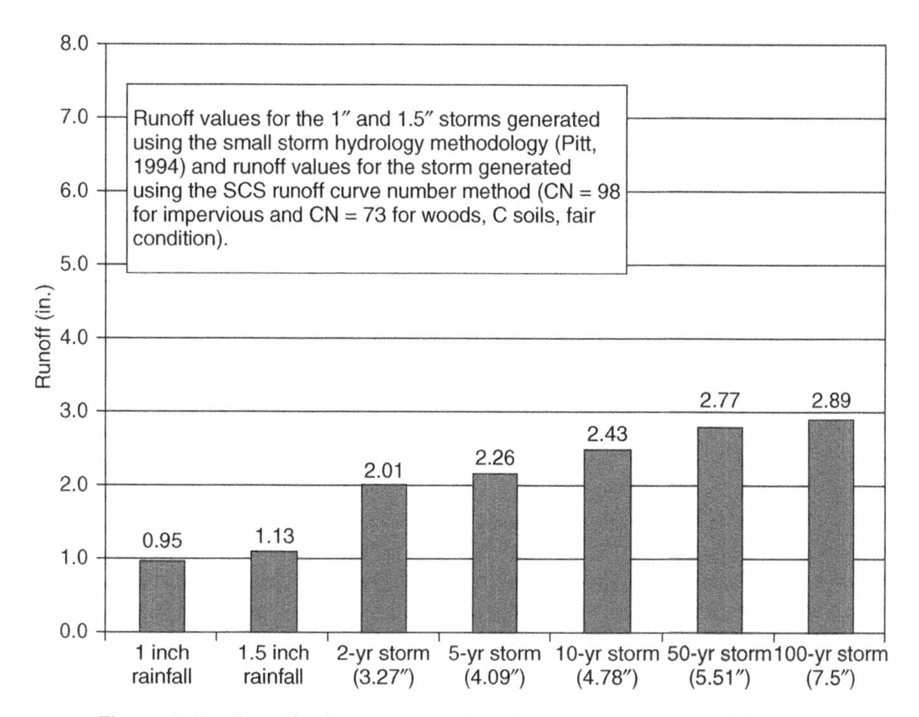

Figure 1-19 Runoff volume increase from impervious cover on HSG C soil.

soil, while the better-drained (B) soil produces a 2.60-in. runoff volume increase. Thus, a volume control guideline must be based on the net change in runoff volume for a given frequency rainfall, in order to be equitable throughout the country on any given development site.

When the balance of a developed site is cleared of existing vegetation, graded, and recompacted, it also produces an increase in runoff volume. Traditionally, if the original vegetation were replaced with "natural vegetation" such as a lawn, the runoff characteristics would be considered to be equivalent to the original natural vegetation, but the fact is that the disturbance and compaction destroy the permeability of the natural soil and increase soil density, which has a direct impact on permeability. *Lawnscapes* actually produce significant runoff during rainfall, laden with nutrients, pesticides, and herbicides.

Consideration of runoff volume control has focused on the frequent rainfalls that comprise a major portion of events in most parts of the country. Since the 2-year rainfall comprises some 95% of the total rainfall in most regions, control of the 2-year event becomes the recommended basis for stormwater management as a *control guideline*. It is considered unreasonable to design any stormwater volume LID measure for greater than a 2-year event. The increase in runoff volume from the 100-year rainfall after site development is so large that it is impractical to require management of this total increase in volume. During such extreme events, the cumulative impact of this volume from largely impervious parcels and in developed watersheds simply overwhelms the natural and human-made conveyance elements of pipes and stream channels. In practice, a best management practice sized for the increase in the 100-year runoff volume would be empty most of the time and would have a 1% probability of functioning at capacity in any one year. Of course, large storms still need to be managed in terms of flooding and peak rate control, to the extent possible. If the 2-year frequency runoff increase can effectively be removed from runoff during the 100-year event, the increase in the peak of the flow rate can be mitigated.

1.5 THE IMPACTS OF DEVELOPMENT ON THE HYDROLOGIC CYCLE

The impact of stormwater runoff from land development on surface water systems such as rivers, lakes, and coastal estuaries is direct and obvious, and the turbid storm flow observed in every small and large watershed that has experienced substantial development is commonplace across the country. What is not so obvious, however, is the impact of land development on the subsurface or underground elements of the hydrologic cycle, as illustrated in Figure 1-2. Most of the rainfall that occurs in any part of the country immediately soaks into the soil mantle under normal climate conditions and is subsequently utilized by surface vegetation in photosynthesis, returning almost half of the annual precipitation back to the atmosphere. The landscapes that are covered with established woodlands and forests are the most efficient in this process, and every tree can be considered a "water pump," drawing moisture from the soil mantle. Where surface vegetation is comprised of less efficient systems, such as grasses or even cultivated crops, the process still comprises a major fraction of the total rainfall. While the amount of evapotranspiration (ET) expected from different types of vegetative systems across the country and the fraction of annual rainfall that this ET represents vary by physiographic region and season, the hydrograph separation of stream flows [17] remains the best method to observe how the runoff process takes place in a watershed over a given year.

Reduced Groundwater Recharge

Under natural conditions, most rain soaks into the soil mantle most of the time. When we cover the soil surface with impervious material and compact the balance of a land development parcel, we effectively prevent the infiltration process from recharging the groundwater aquifer system. The rainfall that would have slowly percolated into the soil mantle and weathered bedrock can no longer reach the *water table*, or *zone of saturation*. This loss of recharge is difficult to quantify, because the only real measurement of this phase of the hydrologic cycle is the reduction in stream base flow as a watershed undergoes the transformation from woodland to cultivated land to suburban or urban land cover. Our records of continuous stream flow on urban tributaries over time can identify the change in flow regime with urbanization, with dramatic increases in storm runoff volume and rate followed by a base flow that frequently approaches zero, altering and frequently destroying the aquatic habitat.

Reduced Stream Base Flow

Because the mechanism of infiltration is not observable on the land surface, we can only measure it indirectly. The same observations of stream and river flow that we use to quantify the discharge in a surface channel serve to measure the groundwater as a part of this flow. When it has not rained in a watershed for several days, all of the water observed (and measured) flowing in the stream (Figure 1-20) or surface channel is coming from the subsurface storage system. During an extended period of stream flow measurement, the flow observed can be distinguished between that which occurs immediately following rainfall as direct surface runoff, and that discharge which comprises the stream flow during the balance of the year. This *base flow* usually represents the major portion of surface stream flow during most of the year, usually on the order of 60% of the annual volume and comprising stream flow about 90% of the year.

Within a given physiographic region, where the annual rainfall is relatively uniform, the base flow can vary significantly. This variation is generally attributed to the surface vegetative system, which can utilize more or less of the infiltrating rainfall. For example, the water balance can vary greatly in different watersheds across the country. Climate also plays a major role, with longer summer seasons and higher average temperatures producing greater evapotranspiration.

Countless small (and large) perennial streams in urban portions of the country have undergone a slow transformation over the past century to a channel that is frequently dry or experiences a very low base flow most of the year, and then is flooded during relatively minor rainfalls as the rapid runoff from extensive

Figure 1-20 Base flow in a perennial stream occurs 90% of the time.

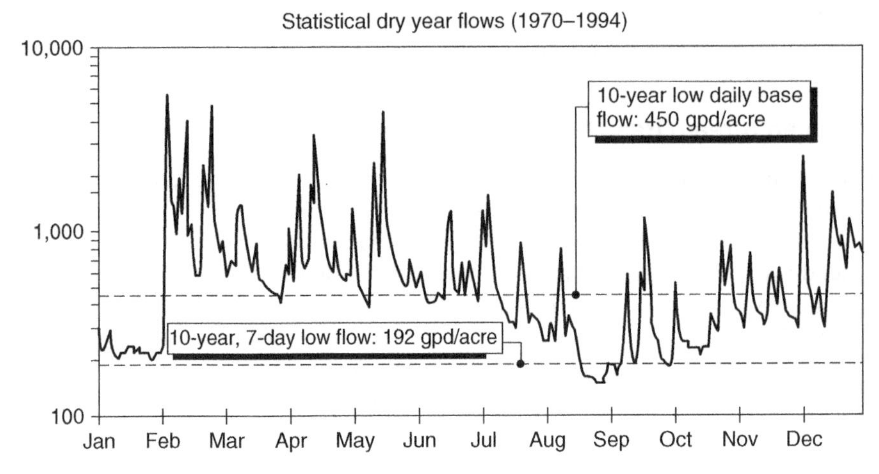

Figure 1-21 Base flow analysis of minimum low flow in a stream.

impervious surfaces pours into the stream. The reduction in base flow of any surface stream usually represents a significant loss of water quality and aquatic habitat, as countless studies have documented. In effect, when we reduce the base flow below the minimum required to sustain the aquatic habitat (Figure 1-21), the system dies or is transformed into something very different.

Altered Stream Channel Morphology

While the increased runoff volume produced and the pollutant load transported with stormwater are important, the impact of this flow in the network of natural stream channels through which it passes is of equal concern. Even where a detention system has been constructed in one or many development sites, the total increase in runoff volume results in a greater flow for longer periods, which affects the natural stream channel by altering the morphology of these channels. Since most surface streams have reached equilibrium with the flow variability under natural conditions, the shape, size, depth, and bank condition of the riparian section is long established prior to land development, although larger river systems that have formed a broad floodplain over millennia will move the channel within this floodplain as discharges vary [18]. On smaller streams that have no tributaries, referred to as *first-order streams*, the impact of the increased volume of runoff is significant, and erodes the channel banks to increase the non-point source pollutant load as the channel readjusts to convey this greater flow.

Water Supply Impacts

A modern water treatment plant is capable of removing virtually all of the sediment and associated pollutants that are attached to the soil particles from a water source. Trace pollutants that are soluble in the raw water are more difficult to remove. The disinfection processes that are a part of this treatment are also capable of removing or killing all of the disease-producing microorganisms found in rivers and lakes. Where only disinfection is included in the treatment process (such as the City of New York), the potential for infectious protozoans, such as *Cryptosporidium parvum* and *Giardia lamblia* to survive the disinfection treatment is very real and can pose a significant health threat, especially to a compromised population (e.g., HIV-infected citizens). The initial sources of these protozoans are wild and domesticated animals, but infected human wastes can also transmit them with runoff.

Perhaps the greatest impact of degraded streams on potable water supply sources is the subtle taste and odor impacts produced in a water supply, especially during warm summer months when raw water reservoirs may themselves be heavily laden with algae. Again, it is the phosphorus conveyed with stormwater runoff that creates the eutrophic environment in the raw water reservoir. In some developed watersheds, surface supply intakes are suspended during periods of heavy rain, but this is not always possible in many supply systems.

Although most modern water treatment plants are very efficient, they do not remove every pollutant conveyed in stormwater runoff, and an increasing number of pollutants are being found in our water supplies. The long-term solution, of course, is to prevent or eliminate their use in the environment, but many are decomposition or end products of common and widely used materials that have become a part of our living environment and cannot be eliminated easily.

1.6 THE HISTORIC APPROACH: DETENTION SYSTEM DESIGN

In the early 1970s, as we began to recognize the hydrologic impacts of land development, the immediate concern was the condition of natural drainage immediately downstream of the building development, and this impact was perceived as a flow-rate problem. We took the technology developed several decades previously for small earthen dams built in agricultural sites (the farm pond) and tailored it to serve as a rate reduction system for the increased runoff volume produced by new impervious surfaces and soil alteration. The detention basin quickly became the design standard, in both a technical and regulatory sense, for stormwater management at every new development site across the country, with significant site impacts (Figure 1-22).

The design criterion became *flow-rate control*, and the stated goal of every stormwater management system was to assure that the rate of stormwater runoff generated from a building site was no greater following development than occurred prior to development. Initially, detention structures were designed primarily to control the extreme runoff conditions of the most severe rainfalls, once again applying the river structure guideline of the 100-year-frequency storm to a small earthen basin built on the upland or in some small drainage swale. With experience, it was recognized that smaller rainfall events could also

Figure 1-22 Detention basin construction removes vegetation and compacts the soil, reducing infiltration and evapotranspiration.

produce erosion impacts downstream, so the detention basin outlet structures were modified to prevent any increase in rate for more frequent storms, such as the 2-year event. The size of any detention basin, however, was based on mitigating the peak rate of the 100-year-frequency rainfall.

As the regulatory programs and design guidelines evolved during the 1980s, it became apparent that a stormwater management system limited to flow-rate mitigation by detention was inadequate. The dramatic increase in runoff volume from developed landscapes was recognized as an important but largely unmitigated result of land development. While the standard detention basin might allow a limited amount of infiltration into the soil mantle at the bottom of the basin, compaction during construction virtually eliminates this possibility. Over time, some designs allowed vegetation to be established within the basin, and this could provide evapotranspiration during and following impoundment, but the actual reduction in runoff volume afforded by the standard detention basin was very little. For the most part, runoff detained in this type of structure simply passes through the basin in a matter of hours and adds to the increase in watershed runoff. Sedimentation of suspended solids does occur, with accumulation in the basin bottom, but the smallest particles (colloids) remain in suspension and pass downstream, transporting most of the phosphorus in the runoff as well as synthetic organics.

Figure 1-23 illustrates how a detention basin operates in terms of runoff volume produced from a development site. The increased runoff volume is held in the structure and subsequently released downstream with no significant mitigation of

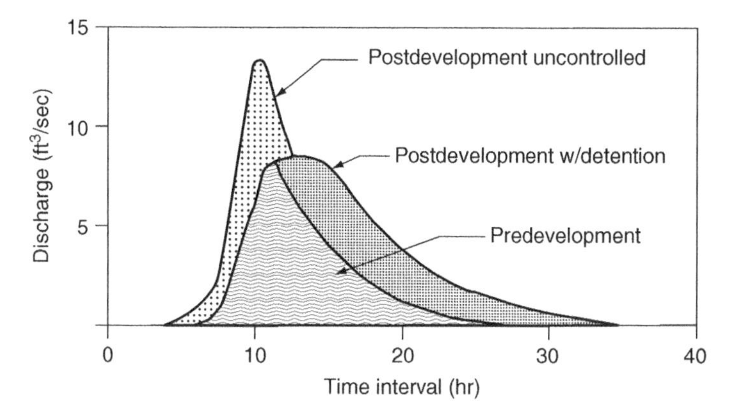

Figure 1-23 Detention basins hold the increased runoff but do not reduce the volume, which subsequently floods downstream.

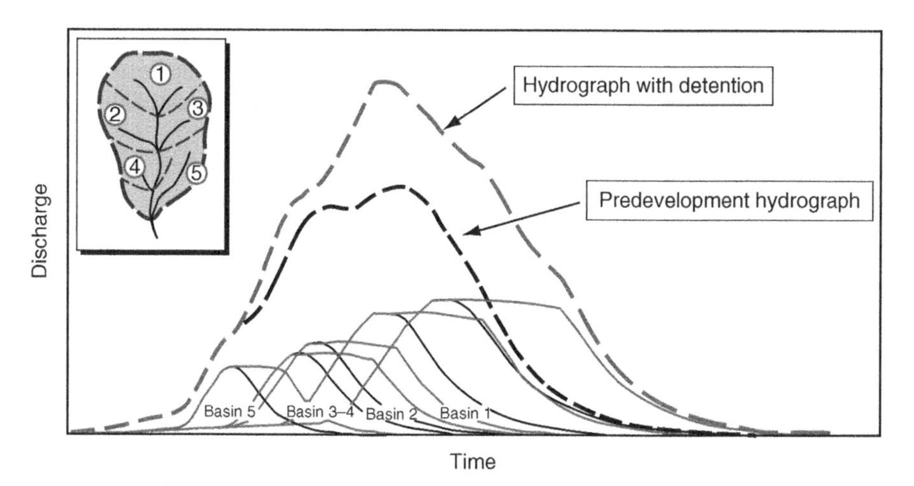

Figure 1-24 Multiple hydrographs in a watershed with detention basins. The compound hydrograph is actually greater than it was predevelopment.

volume. Figure 1-24 illustrates the effect of multiple detention structures within a common drainage basin, with the net result an actual increase in the flow rate downstream, because of the delayed timing in releases from multiple structures. It is obvious that detention-based designs will never be capable of controlling or mitigating the volume increase of rainfall runoff produced by land development. However, we now have countless thousands of these rate control structures situated across the suburban landscape. One of the current issues in stormwater management is how to modify these structures to improve their function beyond simple rate control, as discussed in a later chapter.

1.7 STORMWATER VOLUME METHODOLOGIES

The land development process results in significantly greater volumes of runoff and conveys land pollutants to surface waters, but the difficult issue remains as to how to prevent or reduce this impact. Before we consider how best to "manage" our runoff, we must decide how much of the net increase should or can be reduced or prevented. Since our measurements of the hydrologic cycle are limited to input (rainfall) and output (stream flow), understanding what happens in between in a watershed or catchment is the initial issue.

This is certainly not a new problem, and engineers and hydrologists have developed a number of analytical methods over the past several decades to estimate the amount of runoff produced in a watershed. These various methods are based on the land that comprises the drainage area, and attempt to replicate the complex process of how surface runoff is produced and the multiple pathways followed by each raindrop as it follows the energy gradient downhill. Whatever the algorithm formulated to describe the process, the end result is to estimate the form of the resulting hydrograph that occurs in the receiving stream following rainfall. Since this surface flow hydrograph is the end result of the process, all models "calibrate" or adjust input parameters to replicate this energy waveform.

Since most of this effort has been undertaken to allow the building of structures within the stream or river channel, in the form of culverts, bridges, dams, and other hydraulic structures, the key value measured (or estimated) by the hydrologic modeling analysis has been the *peak rate* of flow that will result during a given rainfall at a specific point in the drainage system. The current stormwater management strategy of reducing or mitigating runoff volume produces a very different perspective on these various modeling procedures, and begs the question: Which method best estimates the change in runoff *volume* resulting from land development?

Many methodologies have been developed to estimate the total runoff volume, the peak rate of runoff, and the stream hydrograph produced from land surfaces under a variety of conditions. In Chapter 6 we describe some of the methods that are most widely used throughout the country. It is certainly not a complete list of procedures, nor is it intended to discourage the use of new and better methods as they become available. There are a wide variety of both public- and private-domain computer models available for performing stormwater calculations, and these models use one or more calculation methodologies to estimate runoff characteristics, following procedures discussed in Chapter 6. To facilitate a consistent and organized presentation of information throughout the country, to assist design engineers in meeting the recommended control guidelines, and to help reviewers analyze project data, a series of worksheets are provided in Appendix A for design professionals to complete and submit with their development applications.

REFERENCES

1. National Oceanic and Atmospheric Administration, 2010. Climatology Records, NOAA, Department of the Interior, Washington, DC.
2. Bailey, R., 2011. Photosynthesis. http://biology.About.com/od/plantbiology/.
3. Fagan, B., Ed. 2009. *The Complete Ice Age: How Climate Change Shaped the World*. Thames & Hudson, New York.
4. Ochoa, G., J. Hoffman, and T. Tin, 2008. *Climate: The Force That Shapes Our World and the Future of Life on Earth*. Rodale Publishing, Emmaus, PA.
5. Ibeqbuna, O. E., 2010. The Importance of Trees in Our Environment. http://ezinearticles.com.
6. U.S. Department of Agriculture Forest Service, 2004. *The Value of Trees*. Urban and Community Forestry Appreciation Tool Kit, NA-IN-02-04. USDA Forest Service, Washington, DC.
7. Yellowstone Caldera, Wyoming. Yellowstone Hot Spot. http://vulcan.wr..usgs.gov/Volcanos/Yellowstone/description.
8. Running out of water. *Scientific American*, Aug. 2008.
9. Benke, A. C., and C. E. Cushings, Eds., 2005. *Rivers of North America*. Elsevier Press, Burlington, MA.
10. 1,000-year storm leaves wreck behind, Flooding of the Rancocas Creek, Burlington County, NJ. *Philadelphia Inquirer*, July 14, 2004.
11. U.S. Department of Agriculture Soil Conservation Service, 1972. *National Engineering Manual*, Chap. 14. USDA SCS (NRCS), Washington, DC.
12. California Environmental Protection Agency, State Water Resources Control Board. http://www.waterboards.ca.gov/water_issues/programs/low_impact_development/.
13. Deru, M., 2008. *Building Design and Performance*. U.S. Department of Energy, National Renewable Energy Labs, Boulder, CO.
14. Green, D., 2007. *Managing Water: Avoiding Crisis in California*. University of California Press, Davis, CA.
15. Schuler, T., and H. Holland, Eds., 2000. *The Practice of Watershed Protection: Techniques for Protecting Our Nation's Streams, Lakes, Rivers and Estuaries*. Center for Watershed Protection, Silver Spring, MD.
16. Cahill Associates and Andropogon Associates for the Facilities Department, Univeristy of North Carolina, 2005. *Stormwater Management Report*. UNC, Chapel Hill, NC.
17. Sloto, R., and M. Crouse, 1996. *HYSEP: A Computer Program for Streamflow Hydrograph Separation*. WRIR 96–4040. U.S. Geological Survey, Washington, DC.
18. Leopold, L. B., M. G. Wolman, and J. P. Miller, 1985. *Fluvial Processes in Geomorphology*. Dover Publications, New York.

Additional Source

Cascadia, 2009. *The Living Building Challenge*. Cascadia Chapter, U.S. Green Building Council, Seattle, WA. http://www.cascadiagbc.org.

2

STORMWATER HYDROLOGY AND QUALITY

2.1 OVERLAND FLOW: THE BEGINNING OF RUNOFF

Many excellent textbooks explain and explore the subject of hydrology [1, 2], and other references discuss the chemistry of pollutants conveyed in stormwater runoff [3, 4]. The suspension, dissolution, and transport of pollutants in stormwater are complex processes that are best understood by considering a few simple questions. The first question is: How does surface runoff begin? This process is considered the "hydrologic response" in a given catchment, watershed, or drainage basin, and the transport of surface pollutants which are scoured from that drainage area and conveyed downstream starts with surface runoff.

The physical model of this process under natural conditions was illustrated in Figure 1-2 with a surface that is vegetated in woodland on a natural soil mantle of well-drained material overlaying bedrock with a fairly shallow water table (10 to 15 ft below the surface). At the initiation of precipitation, the rate of rainfall follows the temporal pattern shown in Figure 1-9. When the rate of rainfall that reaches the soil surface exceeds the rate of infiltration, the surface becomes saturated. Then the pull of gravity begins to move the surface moisture down-gradient across the land surface, following the path of least resistance. This fluid movement is by no means uniform and is discontinuous, with numerous "nooks and crannies" serving as intermediate storage areas. In effect, the flow of rainfall

Low Impact Development and Sustainable Stormwater Management, First Edition. Thomas H. Cahill.

runoff across a natural landscape is a process of interconnected microchannels conveying runoff between small depressions in the land surface, and does not approximate the concept of *overland flow* utilized in many hydrologic models. In that analogy, a *friction factor* is assumed to be applied to the vegetated surface, and the runoff is assumed to be a sheet of uniformly deep flow moving across the landscape. In reality, the movement of surface runoff on a natural landscape is far too complex to describe with a single equation.

Some investigators have also suggested that the process of rainfall runoff from natural land surfaces varies both spatially and temporally as the distance from perennial channels increases. Described as *partial area hydrology* [5], this concept proposes that the initial runoff volume is generated from land surfaces within the floodplain or within the drainage network where wetter soils exist. As precipitation increases during a storm, the runoff area expands in an ever-expanding pattern along waterways and is by no means uniform throughout the watershed. The problem with this analogy is that every watershed would need to be described in terms of the spatial variability of *hydrologic soil groups*, and the runoff produced would increase as the area of runoff expanded during rainfall. No computer model of this mechanism has yet been formulated, but the concept is important when considering how some portions of a watershed may contribute greater pollutant loads to runoff.

During rainfall, water is retained in many soils for extended periods and as surface conditions change, may ultimately be removed by deep roots or even evaporation. Agricultural productivity has been monitored for decades by modeling the soil moisture content as a guide to drought conditions. However, the fraction of rainfall that ultimately reaches the zone of saturation, or water table, is fairly constant from year to year in any given physiographic region.

Since we can measure the aquifer recharge as base flow only when it is finally discharged to a surface stream, it is difficult is gain direct insight into the intermediate subsurface movement of rainfall as "groundwater." We can develop networks of wells and measure the changing depth in the gradually fluctuating water table, and we can withdraw water from a single well and measure the water table change in nearby wells, thereby measuring implicitly the movement through the aquifers [6]. This tells us a great deal about the permeability or *transmissivity* of the rock itself, but does not adequately measure the movement of water through the three-dimensional saturated matrix of rocks. Several computer models have been developed [7] that simulate this movement by considering the rock to exist in layers, each with different permeability, then calculating the water movement from a unit volume in six directions. Although this type of simulation has produced some good approximations, especially on larger scales, the overall movement of infiltrating rain through the subsurface remains a complex process, with gravity as the fundamental energy of movement.

In theory, a continuous recording stream gage situated at or near the mouth of a river or stream will measure all of the water draining out of the watershed. It is possible to separate the volume of flow measured into the two contributing sources: base flow from the groundwater discharge and surface runoff immediately flowing precipitation. In most undeveloped perennial streams and rivers, the base flow comprises about 60% of the annual discharge and occurs about 90% of the time; that is, when it has not rained in a day or two, all of the stream discharge is comprised of base flow.

2.2 REGIONAL HYDROLOGY

The landform, or physiography, creates the boundaries between watersheds and controls the movement and direction of surface waters and rainfall runoff. The amount of runoff that occurs within a basin in any given year, however, is the result of regional hydrology and climate. Most of the United States experiences seasonal differences because of location, and within any given climate zone fairly large variations are possible from year to year. In northern watersheds, the available water can be stored in the form of snow for many months before it is released to the surface stream network, resulting in dry conditions in a watershed even when rainfall (or snowfall) has occurred. While the hydrologic cycle remains a constant over time, the amount of infiltration, evapotranspiration, surface runoff, and base flow can vary greatly with relatively equal rainfalls.

Historically, water was presumed to be a resource to be exploited, and a watershed was viewed as capable of "yielding" the total incident rainfall. A few decades ago, some studies proposed that vegetation be totally removed or greatly reduced in heavily wooded basins in North Carolina to eliminate evapotranspiration (usually more than half of the annual rainfall) and allow greater runoff capture in lakes and reservoirs for human activities such as cultivation and consumption. With an increased awareness of the importance of woodlands and forests in sustaining not only the hydrologic cycle but also the atmospheric balance of CO_2, these types of schemes have faded from general consciousness, although in some parts of the world they are still given consideration.

In any region of the country, the cycles of extreme weather, which from a water perspective translate into flood and drought, are exacerbated by human activity. This is especially true in those basins or regions that have experienced extensive development and the construction of impervious surfaces, because every drop of rainfall becomes immediate polluted runoff. In many highly developed watersheds, the total discharge of surface waters may remain the same, but the distribution between immediate runoff and base flow changes dramatically, as does the ecology of the receiving streams and rivers.

In Chapter 1 we offered some statistics on rainfall variability in different regions of the United States, but perhaps the most telling statistic is the difference between rainfall and evapotransporation for very different parts of the United States. This is the volume of rainfall that is available within a given watershed for all uses, both natural and human. It also illustrates the seasonal variability of the two major elements of the hydrologic cycle and the resulting surface flows that support or limit land development.

Wetlands

There are many potential land development sites that are constrained by wetlands and other sites where an existing wetland system largely prevents development or requires the destruction of the wetland in order to build. The most important concept to grasp in considering this site condition is that wetlands do not exist in a disconnected manner from the balance of the site or watershed, but are a reflection of natural surface or subsurface drainage conditions that produce a saturated land surface for at least a portion of the year. In simple terms, they generally are places where the surface–groundwater connection is most apparent, and in many ecosystems where rainfall is emerging from the groundwater.

There can also be geologic conditions that result in closed surface depressions where wetlands form on the surface, especially in regions of the country that experienced glaciers during the last Ice Age (ending about 25,000 years ago), where the receding glaciers (Figure 2-1) left behind blocks of ice to form "kettle holes" or other depressions in the landscape. The landscapes of the upper Midwest, in states such as Minnesota and Wisconsin, contain rich ecologies of wetlands that are internally drained and reflect the regional water table without a surface outlet.

For most watersheds, however, the wetlands that have formed along the riparian stream or river channels, or around springs and seeps on the hillside, are locations of groundwater discharge. Regardless of the hydrogeologic origins of a wetland, they provide an essential component to the water balance, and in many ways are the beginnings of the land–water continuum. They are an essential part of the hydrologic cycle, and their loss affects the entire watershed.

Figure 2-1 The last ice age came to an end about 25,000 years ago.

Figure 2-2 This highly impervious watershed (90%) has even paved the stream bank (Darby Creek, Pennsylvania).

After decades of destroying natural wetlands, generally in the name of cultivation, we have come to recognize and appreciate their ecologic value. The statistics on wetland filling and draining processes, some of which are still being carried out, are depressing from an ecological perspective. Niering and Littlehales [8] summarize the loss of this resource across North America during the past three centuries, and also beautifully illustrate regional systems. The impact of these losses is much more significant in an individual watershed that has been "urbanized" during that same period. There are countless examples surrounding each of our urban centers, but the Darby Creek in suburban Philadelphia is a prime example of the "paving over" of a watershed (Figure 2-2), where scarcely a wetland can be found except at the lower end of the basin (John Heinz Wetland Center, USF&WS, Tinicum, PA). Here the ground was simply too wet to be built upon without massive filling, and while conservation efforts have preserved some of the original wetland habitats, little remains of the natural system.

The question of greatest interest here is: How do we develop a site that contains wetlands, or is it impossible to do so without further loss? The answer to this question turns on the specific site drainage conditions, since most wetlands occur in the lower-lying portions of a tract of land, but are usually sustained by groundwater flows that begin on the surrounding uplands, the very areas that would be considered "high and dry" prime development land. Where this is the case, the types of infiltration and groundwater recharge of rainfall systems discussed here must be situated at or near the built structures, to assure the continued recharge of groundwater. One critical aspect of most wetland systems is that they are supported by a shallow water table, and upland recharge must maintain the water table elevation in the wetland portion of the parcel. Countless wetlands have been destroyed by a relatively small lowering of the local water

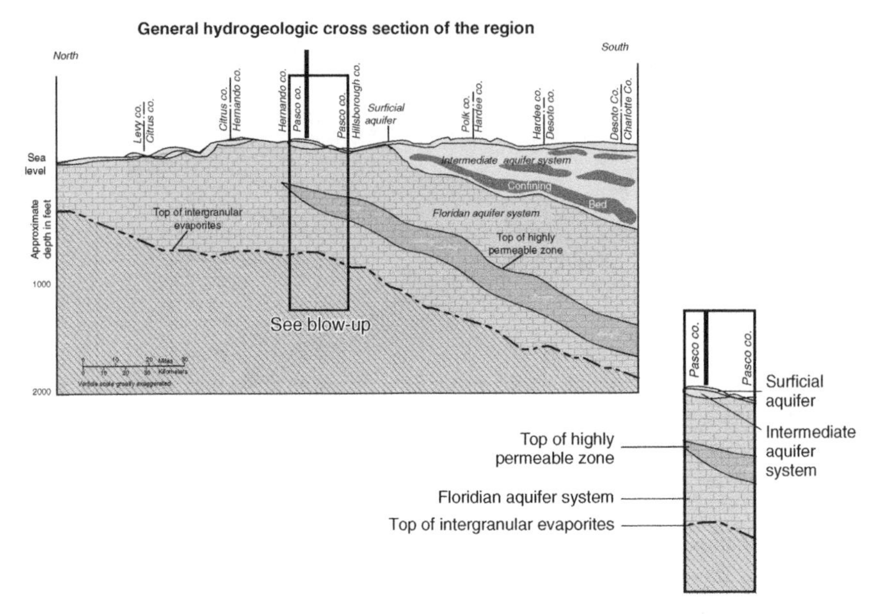

Figure 2-3 Florida hydrology is affected by a slight lowering of the water table, so recharge is critical.

table, especially in very wet environments such as Florida (Figure 2-3). The lowering of the shallow water table is the result of both upland impervious surfaces and substantial groundwater withdrawals for potable supply or irrigation.

First-Order Streams

The emergence of groundwater from seeps and springs represents the continued flow of infiltrated rainfall across the land surface, producing what we refer to as *first-order streams* (Figure 2-4). This term defines a single perennial (year-round) stream that flows downgradient to combine with ever larger elements of a watershed or drainage network. The concept of surface stream ordering [9] is one that has received great study in the analysis of fluvial processes in surface streams, but defines much more of stream hierarchy from an ecological perspective.

To consider a first-order stream as the beginning of life in a watershed is not an exaggeration. The groundwater emerging from the ground is rich in dissolved minerals but does not contain any living organisms, except bacteria from the soil mantle. The exposure of the flowing water to sunlight, oxygen, and the detritus of decaying surface vegetation begins a food chain that quickly evolves from simple decomposing microbes to higher life forms, ultimately resulting in the finfish and shellfish that we have long harvested as a protein source.

Figure 2-4 First-order streams begin as seeps and springs and are the most vulnerable element of the hydrologic cycle.

This complex web of life that begins in small surface streams that ultimately flow into and form mighty rivers and estuaries is an important consideration in the land development process. When we recognize our "watershed address," and the sensitivity of the land–water system in which we plan to develop, we will better understand the interdependency between land and water resources.

During our brief three centuries in North America, we have literally moved land development from the mouths of major river systems to the source of those systems in the headwaters, where countless first-order streams sustain the entire drainage system. The interdependency is obvious if we think carefully about where we build, but past experience indicates that this concern has not been given voice until very recently.

2.3 STORMWATER VOLUME

Once the water balance has been estimated for a given site, the guidelines for sustaining this balance must be established. These technical guidelines evolve into specific design criteria that can be applied to a broad spectrum of stormwater management solutions, commonly referred to as *best management practices* (BMPs), which have evolved as the tools for LID. This book presents stormwater management principles and recommends site control guidelines for volume, water quality, and rate. These guidelines are proposed as the basis for municipal stormwater regulation, and offer guidance for municipalities desiring to improve their stormwater management programs. Some state laws and regulations manage stormwater directly at the state level, while some state-level management occurs through programs such as NPDES Phase II permitting, or the MS4 program, but whatever the regulatory format, the first step is to "set the bar" for our LID strategies.

While any volume control guideline must be quite specific concerning the volume of runoff to be controlled from a development site, it should not limit the methods by which this can be accomplished. The selection of a BMP, or combination of BMPs, is left to the design process. In all instances, minimizing the volume increase from existing and future development is the goal. The LID measures described in this book place emphasis on infiltration of precipitation as an important solution, but this is only one of the three basic methods that reduce the volume of runoff from land development. These three methods can be summarized as follows:

1. Infiltration
2. Vegetation systems that provide evapotranspiration
3. Capture and reuse

All of the stormwater management systems described in this book include one or more of these methods, depending on specific site conditions that constrain stormwater management opportunities.

2.4 THE WATER QUALITY IMPACTS OF LAND DEVELOPMENT

The health of our rivers and streams reflects that of the entire ecosystem in which we live, so their importance to our species goes well beyond the immediate needs of potable supply. Every stream is a living system comprised of many parts, with the web of life critical to our own survival, even though in the United States we may no longer go to the river each day to fish, drink, or bathe. This aquatic system is delicately balanced and supported by the natural flow in the stream, with the largest portion of that supply coming from groundwater discharge, as base flow. Thus, any major reduction in base flow and corresponding increase in polluted surface runoff degrades that surface system not only by runoff discharges but also by reduced base flow (Figure 2-5).

The relationship between rainfall infiltration and subsequent base flow discharge from the subsurface to surface streams is usually not recognized as an important factor in surface water quality, but in fact it is the most important source of water for the aquatic environment. Some research effort has been made to determine exactly how depleted a stream flow can become without affecting the habitat of finfish species. Such models suggest that when the Q7-10 flow (the lowest 7-consecutive day flow with a 10-year frequency) is not sustained, trout are unable to move from pool to pool, and spawning is prevented. Each and every stream has its own set of low-flow impacts, but at some reduction in long-term base flow, the system cannot be sustained and the aquatic community collapses.

As a watershed undergoes urbanization, we can recognize degraded environmental quality beyond sediment-laden streams and polluted lakes. Long-term

Figure 2-5 Reduced base flow has a major water quality impact, destroying the aquatic environment.

measures of the diversity of fish species in rivers clearly shows that as urbanization increases (using impervious cover as a metric), the fish community degrades. When the upland runoff increases in volume with every rainfall runoff period, it affects the aquatic habitat in several ways. The increased sediment eroded from the upland surface and the additional load eroded from the stream channels covers the stream bottom, smothering the habitat for the macroinvertebrate community that provides the food source for all of the higher organisms, including finfish communities. In addition, the sequence of pools and riffles that support the entire food chain in a natural stream is disrupted. Where the vegetation has been cleared along the channel banks and contiguous floodplain by human activity, usually land cultivation or development, the exposure to direct sunlight alters the aquatic habitat and further degrades the ability of many species of organisms to reproduce. The warming of surface runoff conveyed from upland impervious surfaces (Figure 2-6) also adds to the warming of natural streams, and can alter the habitat for sensitive fish species, such as brown trout, to the point that they cannot survive.

Where stormwater discharges into a lake, the influx of phosphorus-laden sediment can enrich and quickly change the trophic state of any impoundment or estuary. An extensive body of literature documents how this beneficial fertilizer on the land becomes a major pollutant in the water, where it does exactly the same thing; it makes plants grow. Unfortunately, in a natural lake or human-made impoundment, this nutrient excess leads to a plant excess, which in turn decomposes within the reservoir late in the summer, depleting the oxygen concentration and resulting in significant fish kills. The decomposition process also produces severe odors and floating mats of detached periphyton, rendering the use of the lake for recreational purposes impractical. It is estimated that 93%

Figure 2-6 Impervious surfaces warm runoff and affect water quality.

Figure 2-7 Over 93% of the lakes in New Jersey are eutrophic, caused by phosphorus in runoff. Most other states suffer the same condition.

of all the lakes in the state of New Jersey are eutrophic or excessively enriched by phosphorus (Figure 2-7), and countless lakeside communities wage a never-ending battle to sustain their lake water quality as land development surrounds the water body. In Minnesota, thousands of lakes have been affected by agriculture and subsequent land development to produce the same change in trophic state, so both regions share the common problem, as do most other lakes across the United States.

Figure 2-8 Non-point source pollutants in the urban landscape are conveyed in street runoff.

Increased Pollutants in Urban Runoff

After the passing of a thundershower, we find both the air and the land surface much cleaner. All of the atmospheric dust and surface organic detritus, from tree leaves to animal wastes, have been washed from the surface and carried from gutters and roadways to surface inlets, where they enter, accumulate, and are forgotten until the sewer no longer flows (Figure 2-8). When these pollutants reemerge downgradient from a storm sewer and discharge to the nearest stream or waterway, we previously considered the problem solved, in terms of our local community, and gave little consideration to our neighbors downstream. The physical and chemical contents of this stormwater vary and are quite different from the steady discharge of wastewaters from our sewage treatment facilities.

It is this basic difference in pollutant discharge between point (municipal and industrial wastewaters) and non-point sources (conveyed in stormwater runoff) that makes the regulatory process so difficult. In many states, following federal guidance, a sort of pollution bookkeeping process, referred to as total maximum daily load (TMDL) analysis, is applied to determine the level of pollutant reduction (by specific chemical) required to achieve the desired level of water quality. Unfortunately, the use of the term *daily* is inadequate to compare the two very different types of mass discharge, with only point sources occurring on a daily basis. A better solution would be to consider the annual loading generated by the two types of sources on an annual basis, prior to selecting the appropriate load reduction requirements. However, this TMDL process currently guides the federal effort to restore water quality in the nation's waters, so the question becomes: How do we remove the runoff-conveyed pollutants moving through a watershed during some 30 days a year?

2.5 THE CHEMISTRY OF URBAN RUNOFF POLLUTION

We have always thought of pollution as the discharge of some noxious fluid from a pipe into a surface stream or lake. In the current environment, virtually all of these point source discharges are associated with, and the end result of, some wastewater treatment process or municipal treatment plant, which in itself represents a major investment of resources in reducing a water quality impact. Although none of these point sources accomplish the complete removal anticipated in our federal legislation of some 38 years ago (the Clean Water Act of 1972), they have resulted in a significant improvement in water quality in many major surface waters.

The effluent from a sewage treatment plant (STP) is measured by the degree of waste reduction or transformation of organic matter, using a parameter such as biochemical oxygen demand (BOD). This measure is the amount of oxygen utilized by the microbial community in breaking down the oxidizable organic matter present in wastewater, primarily bodily wastes but also everything else that we flush down the drain. Current multiprocess STPs are usually designed to achieve at least 85% BOD removal from a "raw" wastewater, with a corresponding reduction in suspended solids. Depending on the current condition of the receiving waters, other pollutants, especially the nutrients nitrate and phosphorus, may also be reduced by additional unit operations. One might ask what happens to the remaining organic matter not broken down in the STP, perhaps 15% or more of the initial waste load. The answer is that the stream system is expected to complete the transformation process from waste matter to food source as organic carbon is recycled through the aquatic environment.

These parameters have little relevance to the pollutants found in urban stormwater. Certainly the organic matter found in runoff can have the same effect on a receiving stream, exerting a demand for oxygen that can lower the level in a stream below the necessary concentration to support desired species. The concentration of dissolved oxygen (DO) in a natural stream varies diurnally, and during warm summer months a shallow perennial stream can experience DO levels well below survival levels for many organisms. A more representative measure of stormwater pollution load is chemical oxygen demand (COD), which uses a strong oxidizing chemical such as potassium dichromate to measure everything in a sample that can be oxidized chemically, including matter that may take a long time in the aquatic environment to complete the process.

The major constituent contained in stormwater runoff is suspended solids (Figure 2-9), comprised largely of sediment but also including the dust, dirt,

Figure 2-9 Sediment-laden runoff following rainfall in a developing watershed.

and human detritus that cover our impervious surfaces. A constant dust-fall from the atmosphere takes place during dry periods, and most of the atmospheric particulate matter is scrubbed by rainfall. Also contained in this surface runoff are all the microbes that are hard at work breaking down the organic constituents found on the urban street, from pet wastes to food materials. These bacteria are also of the same general type that are found in a wastewater effluent, but usually do not present as great a risk of disease transmission as do those in human waste matter.

This does not mean to suggest that stormwater will not pollute a raw water supply source as severely as sewage, but that the health concerns take a somewhat different form. The drippings of petroleum hydrocarbons and metals from our automotive vehicles are hardly what we would consider a suitable constituent in our drinking water supply, nor are the small concentrations of herbicides, pesticides, and other synthetic chemicals that are washed from the landscape with every storm. As our lifestyles have become more complex, so have our residual wastes, and their discharge into our waterways have health implications well beyond our present level of scientific measurement.

One of the more insidious issues at the current time is the formation of "estrogen mimickers" in the environment, and the potential long-term effect of low-level exposure (probably through our water supply) to human health. The decomposition products of several common chemicals can be found everywhere in our modern water supply sources, and have been suspected of reducing the potency of American males over the past six decades.

Impervious surfaces and maintained landscapes generate pollutants that are conveyed in runoff and discharged to surface waters, but the transport process is neither constant nor uniform. The technical solutions developed over the past decades to remove pollutants from our point source wastewaters were based on a

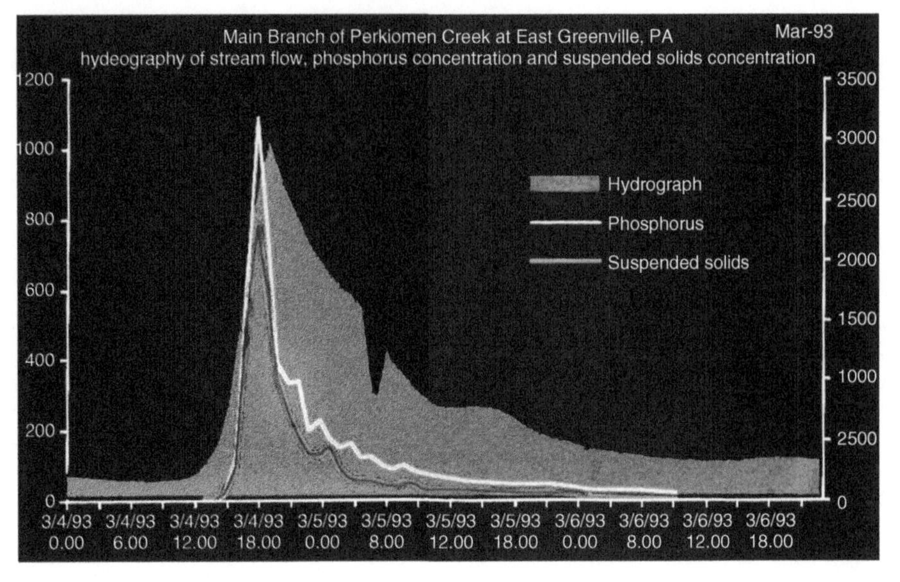

Figure 2-10 Chemograph precedes hydrograph during runoff, requiring double integration to calculate the mass transport.

fairly constant discharge, both in terms of quantity and quality. With stormwater runoff, however, the pollutant concentration for certain pollutants varies by two orders of magnitude during a runoff event (Figure 2-10), and the total volume of runoff generated from a developed site can increase sixfold over predevelopment conditions. When the variability of rainfall is added to this setting, the pollutant removal process requires a different set of technical solutions.

Many of the measurements of pollutant transport in stormwater have documented the fact that a concurrent measurement of chemical concentration over a hydrograph shows a distinct increase that "peaks" before the flow hydrograph, leading to the concept of "first flush" transport. In fact, the particulate-associated pollutants that are initially scoured from the land surface and suspended in the runoff are transported by the kinetic energy of the moving fluid flow and observed in a stream or river before the fluid wave peak that occurs. These pollutants include sediment, phosphorus that is moving with colloids (clay particles), metals, and organic particles or natural detritus. Those pollutants that are dissolved in the runoff (solutes), however, may actually decrease in concentration during heavy runoff as the result of dilution. These include nitrate, salts, and some synthetic organic compounds applied to the land for a variety of purposes.

Managing stormwater to reduce these pollutants includes reducing not only the volume of runoff but also the sources of these pollutants, as well as restoring and protecting the natural systems that are able to remove them from the runoff. These include stream buffers, vegetated systems, and the natural soil mantle, all of which can be used to reduce the discharge of pollutants to waterways.

2.6 UNDERSTANDING POLLUTANT TRANSPORT IN STORMWATER

Our current understanding of pollutant transport in stormwater is summarized next.

Stormwater Quantity and Quality

Water resource qualities are inextricably linked and need to be managed together. Although the most obvious impact of land development is the increased rate and volume of surface runoff, the pollutants transported with this runoff comprise an equally if not more significant impact. In fact, the distinction between water quality and water quantity is imposed and somewhat artificial. Management strategies that address *quantity* will in most cases address *quality*, especially infiltration BMPs.

As land development increases the rate and volume of stormwater runoff, increased runoff scours the land surfaces, both impervious and pervious. The kinetic energy of the raindrops and runoff suspends the solid pollutant particles and transports them with the initial runoff, where the energy is greatest. The pollutants that can dissolve in the rainfall become solutes and move in the runoff in a more distributed concentration. The resulting turbid flow of runoff carries a mix of pollutants, including sediment and organic detritus, nutrients (phosphorus and nitrate), synthetic organic chemicals, and petroleum hydrocarbons. These non-point source (NPS) pollutants generally are materials that are deposited on the land surface. NPS loads are generated in higher quantities from impervious areas that are often defined as "hot spots," such as fueling islands, trash dumpsters, industrial sites, fast-food parking lots, and heavily traveled roadways.

Many so-called "pervious" surfaces, such as chemically maintained lawns and landscaped areas, also add significantly to the NPS load, especially where these areas drain to impervious surfaces (such as gutters) with storm sewers that drain directly to surface waters. The soil compaction process applied to many land development sites results in a vegetated surface that is close to impervious in many instances (Table 2-1) and produces far more runoff than does the original soil mantle. In fact, compaction turns natural soil into concrete. A new lawn surface has probably also been heavily loaded with fertilizers, which results in polluted runoff that degrades all downstream ponds and lakes.

Table 2-1 Increase in Soil Bulk Density with Compaction

Land Use	Soil Density (g/cm^3)
Undisturbed land: forest and woodlands	1.03
Residential neighborhoods	1.69–1.97
Golf courses, parks, athletic fields	1.69–1.97
Concrete	2.2

Source: David B. Friedman, District, Director, Ocean Co. Soil Conservation District.

Some NPS pollutants, such as oxides of sulfur and nitrogen, are even airborne (dust fall), deposited onto the land (or water) surface and then carried into receiving waters, although airborne pollutants in most cases comprise a relatively small fraction of the total NPS pollutant load. Once in a stream, the increased volume and rate of runoff produce additional sediment pollution from streambank erosion by undercutting and resuspension of sediment.

The two physical forms of NPS pollutants are *particulates* and *solutes*. This very important distinction for NPS pollutants is the extent to which pollutants are particulate in form, or dissolved in the runoff as solutes. The best examples of this comparison are the two common fertilizers phosphorus (TP) and nitrate (NO_3-N). Phosphorus typically occurs in particulate form, usually bound to colloidal soil particles. Because of this physical form, stormwater management practices that rely on physical filtering and/or settling out of sediment particles can be quite successful for phosphorus removal, although the phosphorus is bound to the smallest particles (colloids) and may require more extensive removal measures. In stark contrast is nitrate, which tends to occur in highly soluble forms and is unaffected by many of the structural BMPs discussed later. As a consequence, stormwater management approaches for nitrate must be quite different in approach, with wetlands or wet ponds and other biological approaches being more effective, especially where anaerobic conditions can be achieved and where denitrification can occur. Nonstructural BMPs are in fact the best approach for nitrate reduction in runoff, and the easiest BMP is where the surface application of nitrate fertilizer can be reduced or avoided.

Particulates

NPS pollutants that move in association with or attached to particles include total suspended solids (TSS), phosphorus (TP), most organic matter (as estimated by COD), metals, and some herbicides and pesticides. Kinetic energy keeps particulates in suspension, and they do not settle out easily. For example, an extended detention basin is a good way to reduce total suspended solids, but is less successful with TP, because much of the TP load is attached to these small colloids which remain in suspension and pass through detention structures.

Because a major fraction of particulate-associated pollutants is transported with and loosely bound to colloids, their removal by BMPs is especially difficult. These colloids are so small that they do not settle out in a quiescent pool or basin, and remain in suspension for days at a time, passing through a detention basin with the outlet discharge. It is possible to add chemicals to a detention basin to coagulate these colloids to promote settling, but this chemical use turns a natural stream channel or pond into a treatment unit, and subsequent removal of sludge is required. A variety of BMPs have been developed that serve as runoff filters and are designed for installation in storm sewer elements, such as inlets, manholes, or boxes. The potential problem with all measures that attempt to filter stormwater is that they quickly become clogged, especially during a major event. Of course, one could argue that if the filter systems become clogged, they are performing efficiently and removing this particulate material from the runoff. The major problem then with all filtering (and to some extent, settling) measures is that they require substantial maintenance. The more numerous and distributed within the built conveyance system that these BMPs are situated, the greater the removal efficiency, but also the greater the cost for operation and maintenance.

If the concentration of particulate-associated NPS pollutants in storm runoff, such as TSS and TP, is measured in the field during a storm event, a significant increase in pollutant concentration corresponding to but not synchronous with the surface runoff hydrograph is usually observed. This change in pollutant concentration, referred to as a *chemograph*, has contributed to the concept of a *first flush* of NPS pollution. In fact, the actual transport process of NPS pollutants is somewhat more complex than "first flush" would indicate, and has been the subject of numerous technical papers [10–12]. The peak in particulate-associated chemical pollutants precedes the fluid peak because the kinetic energy available moves suspended materials more quickly than the waveform moves in the stream flow, so the recessional portion of the hydrograph conveys only a small fraction of the pollutant load.

To measure accurately the total mass of NPS pollution transported during a given storm event, volume and concentration must be measured simultaneously and a double integration performed to estimate the mass conveyed in a given event. To fully develop an NPS pollutant load for a watershed, a number of storm events must be measured over several years. This technique was implemented in the study of Lake Erie during the late 1970s with respect to the mass transport and loading of phosphorus, the key element in the changes in trophic state that were of great concern [13]. Monitoring of nine major rivers flowing into Lake Erie on a continuous basis allowed investigators to conclude that the land runoff contribution comprised some 60% of the pollutant loading to the lake, and initiated a long-term effort to reduce this input. As has been proven many times since then, the dry weather chemistry is seldom indicative of the wet weather concentrations expected, which can be two or three orders of magnitude greater for particulate-associated pollutants such as phosphorus.

Solutes

The NPS pollutants that are soluble, or dissolve quickly during runoff, generally do not exhibit any increase during storm event runoff, and in fact may exhibit a slight dilution over a given storm hydrograph. Solutes include nitrate, ammonia, salts, organic chemicals, many pesticides and herbicides, and petroleum hydrocarbons (although portions of the hydrocarbons may bind to particulates and be transported with TSS). Regardless, the total mass transport of soluble pollutants is dramatically greater during runoff because of the volume increase. In many watersheds, the storm transport of soluble pollutants can represent a major portion of the total annual discharge for a given pollutant, even though the absolute concentration remains relatively constant. For these soluble pollutants, dry weather sampling can be very useful, and often reflects a steady concentration of soluble pollutants that will be representative of high-flow periods.

Some soluble NPS pollutants can be found in the initial rainfall, especially in regions with significant emissions from fossil-fuel plants. Precipitation serves as a "scrubber" for the atmosphere, removing both fine particulates and gases (NO_x and SO_x), and a pollutant load of atmospheric deposition. Chesapeake Bay scientists have measured rainfall with NO_3 concentrations of 1 to 2 mg/L, which could comprise a significant fraction of the total input to the bay. Other rainfall studies by the National Oceanic and Atmospheric Administration (NOAA) and the U.S. Geological Survey (USGS) have resulted in similar conclusions. Impervious pavements also generate nitrate load, reflecting a mix of deposited sediment, vegetation, animal wastes, and human detritus of many different forms.

REFERENCES

1. Maidment, D. R., Ed., 1993. *Handbook of Hydrology*. McGraw-Hill, New York.
2. Viessman, W., and G. Lewis, 2003. *Introduction to Hydrology*, 5th ed. Pearson Education, Upper Saddle River, NJ.
3. Cahill, T., and T. Hammer, 1976. Phosphate transport in river basins. Presented at the International Joint Commission Fluvial Transport Workshop, Kitchener, Ontario, Canada.
4. Cahill, T. H., 1976. *Modeling Non-Point Source Pollution from the Land Surface*. EPA-600B76-083. U.S. Environmental Research Laboratory, Athens, GA.
5. Dunne, T., and L. Leopold, 1978. *Water in Environmental Planning*. W.H. Freeman, New York.
6. U.S. Geological Survey, 1996. *MODFLOW-96: Transient*. USGS, Washington, DC.
7. U.S. Geological Survey, 2000. *MODFLOW: Three-Dimensional, Finite Difference*. USGS, Washington, DC.
8. Niering, W. and B. Littlehales, 1991. *Wetlands of North America*. Thompson-Grant, Inc., Charlottesville, VA.
9. Leopold, L. B., 1974. *Water: A Primer*. W. H. Freeman, New York.
10. Cahill, T. H., P. Imperato, and F. H. Verhoff, 1974. Evaluation of phosphorus dynamics in a watershed. *Journal of the Environmental Engineering Division, ASCE*, Apr.
11. Baker, D. B., 1976. *Water Quality Studies in the Maumee, Portage, Sandusky and Huron Rivers*. TMACOG Section 208 Study. Toledo Metropolitan Council of Governments, Toledo, OH.
12. Cahill, T. H., 1992. *Limiting Nonpoint Source Pollution from New Development*. New Jersey Department of Environmental Protection, Trenton, NJ.
13. U.S. Army Corps of Engineers, 1977. *Lake Erie Wastewater Management Study*. Phase I Report. USACE, Buffalo, NY.

3

LAND AS THE RESOURCE

3.1 HISTORIC PATTERNS OF LAND DEVELOPMENT

The current process of land development is based on opportunism; a proposed land use is determined by the perception that such a use will thrive in a certain location. The process is driven almost entirely by economic pressures and resource availability, with distinct locations favored. The earliest European settlements, some dating back over 3,000 years (Figure 3-1), were selected based on water availability (Figure 3-2), defensive position, and food sources. Communities were built around a defensive center or castle, and later the defensive structure was expanded to wall in the city.

By the time of the colonial community-building process in the sixteenth century, the defensive position was not critical. New settlements were connected by waterways, roads, and lastly by rail in the nineteenth century, and this became the modern pattern of development in the United States. Until the past century, the village square, marketplace, or town center served all functions, as commercial, government, and cultural center, with early manufacturing distributed throughout the community. People lived, learned, and worked in their part of town or neighborhood, and interactions were pedestrian in nature.

Along the eastern coast of North America, the initial settlements were situated on coastal estuaries and the embayments of major river systems, because

Low Impact Development and Sustainable Stormwater Management, First Edition. Thomas H. Cahill.

Figure 3-1 Tuscan hill town that has been occupied continuously for three millennia.

Figure 3-2 Water supply by aqueduct was added by the Romans 2,000 years ago, allowing continued occupation.

water transport was the only mode possible. The urban centers that evolved over the following three centuries began as communities totally dependent on inland rivers as a source of both potable water supply and commerce. The exploitation of raw materials from the interior of each watershed, first by clearing of woodland and then by cultivation, depended on water transport. As the population moved inland, following the drainage network, raw materials moved downriver to commercial centers arrayed along the coastline, from Boston to Charleston to New Orleans.

The inland waterways served as the primary transportation corridors through the early nineteenth century, when a brief period of canal building augmented this system with connecting elements that paralleled rivers or even crossed watershed boundaries. The river transportation network was enhanced later in the century by numerous locks on the major rivers that allowed bulk materials, such as coal and grain, to move from river center to center. These water highways were soon augmented and in many situations replaced by the rail system that took form in the mid-nineteenth century. This web of rails developed independent of any hydrologic constraints and linked the inland communities with the coastal cities, passing through many watersheds in the process. As the settlement patterns moved westward, the original waterway-based communities along major rivers were connected by rail, eventually binding both coastlines with a transportation grid of steel rather than water. The movement of people and commodities by wagon and horse did not replace the water and rail systems, but it did develop transportation pathways that would form the basis of the present roadway system. The sense of living in a river basin or watershed was quickly lost from the American consciousness.

During the twentieth century in the United States, land planners, architects, engineers, and developers focused their energies on the existing urban centers, and envisioned a future land use pattern that, by and large, was limited to infill with some regional growth along transit corridors. This land development pattern continued through the first quarter of the century (Figure 3-3), but following World War II, the growth patterns literally exploded from the urban core (Figure 3-4), as new families left the cities and relocated to suburbia, where the dream of "open space" and modern living met the reality of commuting, traffic, and the sound of lawn mowers on a summer's morning.

Those who were left behind (or elected to remain in the city) were faced with increasing crime and decreasing property values. As the economic base changed from urban centers surrounded by agricultural lands to a dominance of suburban residential parcels whose occupants work in distant places, the patterns of land development changed. Within the urban center a process of redevelopment was initiated in the late twentieth century to keep the city vital, but as manufacturing and retail moved out to the suburbs, so went the prosperity that had given life to the cities.

The advent of the auto, of course, was the driving force that changed this human dynamic. The patterns of development during the past half century have followed the tire tracks of the highway (Figure 3-5), with residential, commercial,

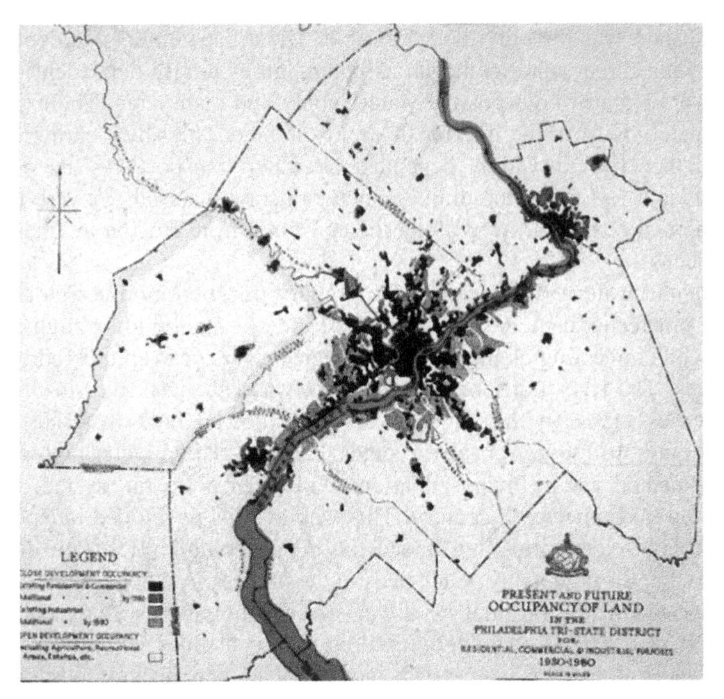

Figure 3-3 Future growth was expected to remain within the urban boundary in most cities (Philadelphia Tri-state District Master Plan, 1932).

industrial, and even recreational uses strung along the road network, repeating the same pattern in every community across the country. The identical strip mall, shopping center, or office park can be found everywhere, frequently with the same mix of retailing ventures. While residential design does vary by region, it must also be recognized that one can find a Cape Cod design, California hillside home, or midwestern ranch house anywhere else in the country. Whatever the particular mix of styles, it is the pattern of repetitive and somewhat boring units marching across the landscape (Figure 3-6) that is most representative of suburban residential land use. Of course, the older urban pattern of "row" homes found in most Eastern cities (Figure 3-7) was equally unimaginative, but did meet a critical need of affordable housing at a time in our development when we could not afford to live far apart or remote from employment or public transit.

The end result of this 50-year-long transition is what we see today: endless tracts of identical residences sprawling across former fields, bound together by an asphalt and concrete web radiating out from the original city center. After several decades of decay, the urban centers have undergone or are just beginning a rebirth of activity, including vertical residences (Figure 3-8) in city centers and along the edge of land and water. Such "vertical villages" serve to induce the return of residents to the city, but as separate communities that are isolated from the less

Figure 3-4 Urban growth expanded well beyond the city into the surrounding countryside following the highways (from the Valley Creek, Chester County Plan, 1990).

fortunate residents who dwell at ground level. This development pattern contrasts with the horizontal sprawl of suburbia and certainly provides residential use with far less space. They are focused on an economic and demographic target, to meet the needs of the aging "baby boomers" who seek to return to the convenience of the urban center.

Figure 3-5 Valley Creek, Chester County (1990). This watershed was in agricultural land use until the highway was built in 1970.

Figure 3-6 Suburban development marching into farm fields has been the pattern of land use change for the past three decades.

Figure 3-7 The pattern of urban development was set a century ago: row homes, paved streets, and mass transit (90% impervious).

Figure 3-8 "Vertical villages" rise along the Seattle, Washington, waterfront.

3.2 SUSTAINABLE SITE DESIGN

The process of sustainable site design begins by distinguishing the location of the proposed site, with an initial classification as "undeveloped" or "redeveloped." In most situations this distinction is easy, with the cultivated or fallow field situated well beyond the town center contrasted with redevelopment in a blighted urban center, although not all situations fit easily in one class or the other. The more important distinction is the presence or availability of utility infrastructure, beginning with road access but also considering water, sewer, and power service, as well as communications. For commercial land uses, the market is determined by the highway connections, their proximity to population centers, and their economic buying capacity.

LID is intended to deal with both land development and redevelopment situations, changing the way we build or rebuild our communities. In this chapter we examine the set of concepts and processes that can begin to restore and sustain the land resources affected by the new land development practices of the past half century, as well as the rebuilding of our cities, a work in progress across the country. The chapters that follow describe specific LID methods and materials, both nonstructural (preventive) and structural, that can be applied to any given site design throughout the country, with consideration of the natural constraints and local hydrology.

3.3 WATERSHED SETTING AND PHYSICAL CONTEXT

Most land development plans are carefully fit within a given parcel, with the building program framed around what is allowed by local land use regulation and the accessibility of the site. Our current paradigm of development is driven by the existing roadway network and usually fails to consider the regional context, especially the hydrologic setting, or watershed. Since the roadway infrastructure is usually the prime determinant in parcel selection, any reference to hydrologic boundaries, such as ridgelines, is unheard of in the current development market.

For this reason alone, sustainable development requires a new perspective. The site design solutions possible will be very dependent on where we are situated within a given watershed, regardless of the relative size. Generally speaking, the headwaters of a given watershed are less likely to support large alterations in topography (grading), more likely to affect the most sensitive elements of the surface drainage system (first-order streams), and result in greater removal of existing vegetation, since the steeper slopes have generally deterred cultivation in historic landscapes.

As the scale of watersheds increase to larger and larger aggregations such as major river basins (the Mississippi or Colorado), estuaries (the Chesapeake or Sacramento) or lacustrine systems such as the Great Lakes, the watershed setting requires a broader perspective. While larger systems usually share a common hydrology and physiography (depending on the scale), each watershed is a separate part of the larger system, and land development that has taken place in each watershed reflects the historic patterns of human habitation.

The cultural environment of the United States in the twenty-first century pays little if any attention to its "watershed address," and most citizens identify their home as living in a local place name, with only a faint understanding of what their watershed address actually means. While the water cycle provides their drinking water, carries away their sewage and runoff, and generally makes their home a livable habitat, it seldom defines a location. The increased understanding and sense of place that is an important part of the transition to sustainable land development is yet to receive general recognition in the public mind, and the evolution of LID should result in a new model of land use planning and site design that recognizes the hydrologic boundaries of every community.

3.4 SMART GROWTH ISSUES

As we refocus our thinking on ways to reduce the impact of impervious surfaces and sprawling land development patterns, the first set of concepts center on clustering our needed development, reducing the land area required and the environmental impact associated with that land use change. Rethinking how and where we build has given rise to several related concepts, with such labels as "smart growth," "new urbanism," "sustainable development," "low-impact development," and "green building." The built forms of these concepts vary greatly and by no means follow a standard formula, but the common elements of limiting the space required, the energy needed, and the resources consumed provide a set of basic design criteria.

One early project that illustrates the site design concepts followed by "new urbanism" is the *Celebration* community in Orlando, build by the Disney Corporation in 1995. As one of the first major projects to follow these concepts, it set the standard for new land planning. In similar but smaller-scale efforts, Figure 3-9 provides a set of illustrations showing prime examples of conservation design applied to land parcels as visual case studies of some well-known developments that are considered representative of the smart growth approach to site design. These projects do not necessarily reflect the best stormwater management design, but they all do an excellent job of minimizing the land required to meet the building program.

Changes Related to Development

Each type of development, from single-family residential through large-scale commercial, will require changes in the standard design model in order to fit the smart growth guidance, but these changes should not deny the designer the opportunity to apply creative concepts in their program. Therein lies the danger in proposing "standards" for any design effort, and reinforces the concept that

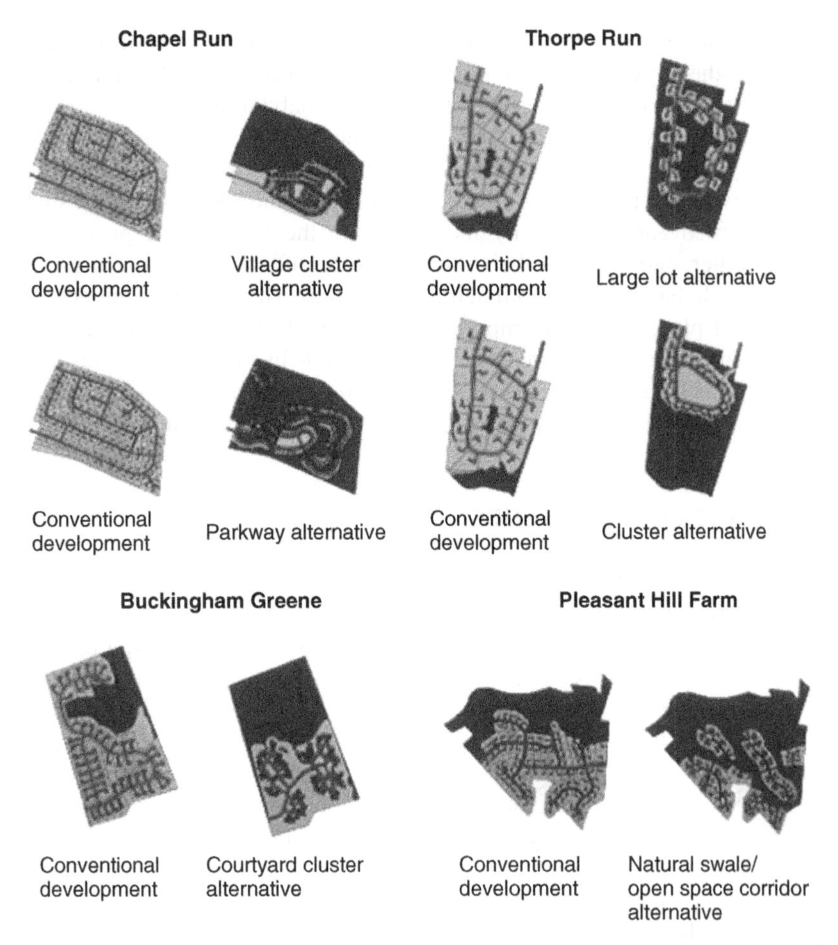

Figure 3-9 Conservation designs for various site plans (Brandywine Conservancy; Wes Horner, 1992).

we must be willing to change our thinking as to what constitutes a sustainable habitat, office, or shopping center. Our traditional approach to land development has been to alter the land to fit the development, and that such change was a normal part of the process. In the sustainable approach, we try to minimize this land alteration, and fit our program (or some portions of it) to the land, taking advantage of the opportunities and constraints that go with the natural landscape.

We also need to rethink the patterns of development. Do we really need to separate all of the activities that comprise our community, with a rigorous set of guidelines that define what can be built where? Although many rural municipalities have not yet adopted a zoning ordinance and map, many if not most of the rapidly growing communities do impose a stringent set of boundaries that separate the settlement into separate (and economically segregated) parts. The ancient village offered a valid model of how to live together without conflict, but was based on a transportation system that was far less intrusive and demanding. The automobile creates a major impediment to the use of the village model for new communities, unless an integration of functions is clearly established and the auto is limited.

3.5 CONFLICT BETWEEN DESIRED LAND USE AND SUSTAINABILITY

When we refer to "desired land uses" we usually mean desired by the owners, developers, or other parties that will profit or benefit from the development. In older communities, especially the larger cities, a prior use may no longer be appropriate to the current situation. Many of the older urban centers situated along inland rivers, such as Pittsburgh, Pennsylvania (Figure 3-10), have riverfront property where commerce and industry is gradually being transformed and replaced by a desire to use the land for mixed office and residential purposes (Figure 3-11). In this example, a "vision plan" has been developed [1] that will accomplish this transition, with two major environmental goals: to restore the hydrologic cycle (by preventing rainfall from drowning the "combined" sewage and stormwater system and producing overflows to the river), and by restoring the tree canopy throughout the entire area, with an ultimate goal of 40% canopy cover. Other cities have set similar redevelopment goals with a large "green" component, but none quite so ambitious. Whatever the growth pressure, be it

Figure 3-10 Pittsburgh, Pennsylvania, at the confluence of the Allegheny and Monongahela rivers, forming the Ohio River, has undergone dramatic land use changes, which continue today, including use of vegetation for stormwater management.

Figure 3-11 A vision plan for the Allegheny riverfront will guide the transformation of land use to LID (see Appendix B).

economic gain or community revitalization, the same rules should apply—build to fit the limits of the land, following sustainable guidelines.

If the land has been severely degraded by prior activity, special restorative measures may be required prior to redevelopment. The isolation of contaminated soils or the removal of materials may be a requirement of preparing a given site for renewed occupation. In many locations it may be impossible to restore the original environmental qualities of a parcel (referred to as the "Humpty Dumpty syndrome"), and the redevelopment process must be more creative in reestablishing a suitable habitat. The redevelopment of the River Rouge plant by Ford Motor Company (Figure 3-12) provided an excellent example in 2002, as they sought to demonstrate the industrial plant model for the twenty-first century. In this $2 billion project, they rebuilt with porous pavement and vegetated roof systems and attempted to remove residual pollutants from the soil with vegetation, on a parcel that had seen a century of contamination.

3.6 PHYSICAL DETERMINANTS OF LAND DEVELOPMENT

Geology

The continent of North America and the middle portion occupied by the United States is a complex geologic region, comprised of very different bedrock regions, with coastal areas underlain be unconsolidated materials, especially along the Eastern and Gulf coasts. The surface geology that underlies the United States was formed over many different geologic periods and by the plate movements

Figure 3-12 The Ford auto plant in Dearborn, Michigan, has become the industrial model for sustainable building, with a vegetated roof, porous pavements, wetland channels, and toxic soil remediation.

that constantly change the surface of our planet, but at a time scale beyond our comprehension. For most of human history, human beings did not occupy North America, and we did not arrive until some 8,000 years ago (as far as we know). Human settlements were certainly guided by and followed geologic constraints and opportunities, but for the most part Native Americans did little to alter the land form or extract mineral resources (Figure 3-13). This changed when Europeans arrived.

The United States can be considered as comprised of major physiographic regions (Table 3-1) that reflect both the geologic formations that underlie and form the land surface and the weathering processes that have shaped this surface over millennia. Much of the middle portion of the United States was a shallow sea during the Cretaceous period (200 million years ago), and the current coastal regions were inundated by ocean waters during periods of sea-level changes caused by warming and cooling cycles. The inland sea resulted in a thick layer of deposited marine organisms that formed chalk (creta), and the other valleys that were covered by ocean waters followed a similar process of marine organism deposition to form carbonate bedrock, or limestone. These rock types have developed solution channels over hundreds of millions of years that hold vast aquifer systems of groundwater, and provide abundant water capacity.

In the balance of the country underlain by older bedrock, groundwater is contained in fractures in the rock and is generally limited to a few hundred feet

Figure 3-13 In the arid southwestern United States, native Americans built their habitats in the walls of canyons near a water source.

Table 3-1 Major Physiographic Regions Within the United States[a]

Eastern	Central	Western
New England/Maritime	Interior Lowland Plateaus	Wyoming Basin
Adirondack	Central Lowland	Colorado Plateaus
Coastal Plain	Ozark Plateaus	Southern Rocky Mts.
Piedmont Plateau	Ouachita	Middle Rocky Mts.
Blue Ridge	Valley and Ridge	Northern Rocky Mts.
Appalachian Plateaus	Great Plains	Columbia Plateau
	Coastal Plain	Basin and Range
		Cascade–Sierra Mts.
		Pacific Border
		Lower California

Source: Data from [Ref. 19, Chapter 1].
[a]Central North America only.

from the land surface. In the wetter portions of the country underlain by the harder bedrock, the discharges from these aquifers as springs or base flow in surface waters has provided a portion of water supply needs, but surface systems were the primary water source until deep-well drilling equipment was developed a few decades ago. Deeper waters are also brought to the surface in geothermal formations, one of which was laid down by a moving "hot spot" or point of discharge from the magma layer beneath the surface plates, such as the source of the geothermal vents in Yellowstone (see Figure 1-6). At other locations on the planet, the geothermal energy can be harnessed and utilized as an excellent

Figure 3-14 Geothermal energy power plants are limited to plate contacts, such as in New Zealand.

alternative to fossil fuels (Figure 3-14), but these geothermal energy sources are limited to geologically active regions along plate contacts.

In deeper aquifers such as the Ogalala formation of the central United States, the stored water is the result of rainfall from some 2 million years ago, while most of the water moving near the surface through fractured bedrock or in solution channels has fallen as rain within the past year or two. For the most part, the subsurface movement of water is confined to a relatively shallow layer of soil and weathered bedrock overlaying more dense formations. In the very deep aquifers, water is highly mineralized and difficult to use for potable supplies.

Physiography

The shape and structure of the land surface determines how and where we occupy it for habitat. In many historic sites, the locations chosen were intended to take advantage of difficult site conditions, rather than convenient access or relatively flat grade. Although the primary determinants of habitat selection for countless decades were a source of drinking water and protection from the elements, we also had to consider protection from each other. Many of our seemingly romantic sites from earlier European history were designed primarily as fortifications along a river, or hillside structures and villages that were positioned to be very difficult to access. The geology and landform were selected for the difficulty of the site rather than the opportunity for land development. Most of these European walled city models were not replicated in the colonial settlements, except for the earliest communities (or forts), and then only briefly. In a democratic society based on collective security, the need for a protective "castle" became irrelevant, although each citizen hopes for his own simple version, for reasons that are intended to impress his neighbors rather than defend against them.

As walled fortifications became unimportant, communities grew up around commercial centers at water, rail, and road confluences, generally on flatter terrain, but still surrounded by open land kept in cultivation to support the center. This city model continued for over two centuries through the agricultural period, with countless small towns covering the American landscape from coast to coast. As our modern cities grew, they expanded out from the urban core into the surrounding farmland, following the rail and primarily the roadway network extending out from the center, as shown in Figure 3-5. In this Philadelphia model, the regional growth anticipated in 1930 grew into the surrounding agricultural counties by 1960, and by 1990 transformed the region into a community with the same basic population, now diffused over five counties and hundreds of square miles. This process was repeated in every major city in the United States, as the economic transformation from a predominantly agrarian society in the nineteenth century to the sprawling urban regions of the twentieth century began a process that seemed to know no end, until the energy reality of the twenty-first century forced a more sustainable model to evolve.

Topography

Of course, since most of the earlier settlements in the United States were intended to optimize the cultivation of land or graze livestock, they were situated in flatter, fertile locations in any given watershed, with water access to coastal or down-river centers of commerce. In recent decades, as we have evolved from a largely agrarian society to an urban industrial culture, the transformation of this open farmland to residential development has characterized many of the issues considered here. In many rapidly developing communities, the availability of open farm fields for conversion to new home communities has been limited, and development has been forced to move into more steeply sloping portions of the landscape. This pressure has resulted in massive loss of woodland in many watersheds and dramatic soil erosion, the result of trying to turn a steep land surface into flat terraces for occupation by roads and houses. The long-term harm created by this alteration of the natural landform is yet to be fully appreciated, although the water quality impact on local stream systems is painfully obvious.

Building on steeply sloping land is considered standard in some regions of the country, not only because of the limited availability of flatter land but also to take advantage of visibility from an elevated site. The southern California model of a home perched precariously on the side of a mesa may be considered dramatic by some and foolhardy by others, but it definitely increases the risk of loss by not fully considering the constraints of topography.

Soil and Subsurface Conditions

The creation of the surface layer of soil from weathered rock is largely due to rainfall, with freezing in cold climates and dissolution of minerals in all climates gradually allowing vegetation to get a foothold and begin the process of making earth. The newest "dirt" is situated at the rock interface, and the oldest earth is at the land surface. It requires thousands of years to form a soil mantle, and the surface is constantly altered by wind and flood, redepositing the weathered soil in distant locations.

The role of the soil mantle as the primary medium for rainfall infiltration was discussed previously, as well as the physical, chemical, and biological processes that remove and break down most of the pollutants that decompose on the surface and are gradually incorporated into the organic surface layer. If the decomposition products or land-applied chemicals can be solubilized, they will move with the infiltrating rainfall and percolate through the soil mantle to the groundwater. Thus, chemicals such as salts, nitrates, and various synthetic solvents and dielectric solutions can become pollutants in the aquifer system. The greatest issue, of course, is when these aquifers provide potable supply sources, and the risk of ingesting these chemicals, even in minute quantities, is a cause for concern.

A natural soil with a vegetated surface responds slowly to precipitation. Initial rainfall collects on vegetation and drips slowly onto the underlying soil surface. Once rainfall has wetted the surface of soil particles (or any porous surface), it begins to drain vertically under the inexorable pull of gravity. This movement is altered by the size of particle spaces, the type of soil, and the variation in particle density or compaction as the rain is pulled downward, with lateral movement comprising a significant component in many complex soils. With the pore spaces open to the atmosphere, the movement is constant, but can vary in duration. The process can take weeks to pass through the soil mantle and become a part of the groundwater system, but it forms a significant component of the hydrologic cycle.

Soil "design" in the land development process seldom considers what lies beneath the surface, other than to attempt a "balance" of required earthwork (excavating and filling) to implement the development desired. This is strictly an economic consideration and reflects the mind set that regards the Earth as a resource to be exploited rather than sustained. It assumes that our perceived needs warrant any alteration deemed necessary to build successfully. In effect, we treat the soil like dirt. One of the greatest obstacles to be overcome in the sustainable design process is to enlighten the development community as to the importance of what lies beneath the ground surface and the Earth itself. This water reservoir storage medium, water purification, system, and growth medium, and its relevance to what we build on the surface, is critical to sustaining the hydrologic cycle.

The best guidance that can be applied to make a development program sustainable is to fit the building program to the land surface, minimizing land disturbance by carefully sculpting the land rather than destroying it. Numerous projects that followed this guidance are discussed or referenced in later sections, but the principle is uncomplicated, and while the solutions will vary in different physiographic regions of the country, the implementation process will evolve through more considerate design efforts that shape the development to fit the land rather than the current opposite practice.

3.7 URBAN COMMUNITIES WITH COMBINED SEWER OVERFLOWS

In many older cities, especially along the East coast but not limited to the East, the sewer system consists of a single pipe, in which both sewerage and stormwater are "combined," an unfortunate consequence of historic development patterns that took shape during the late nineteenth century (Figure 3-15). During periods of rainfall, these sewage systems overflow to surface waters when the hydraulic capacity of the conveyance system is exceeded, resulting in one of the most serious impacts on water quality of the twenty-first century. The discharge from these overloaded sewers occurs at outlets described as combined sewer overflows (CSOs) and is a legacy of urban expansion. From New York and Philadelphia to Chicago and Kansas City and as far west as Portland and Seattle, the issue overwhelms the redevelopment process in urban environments and has created a multi-billion-dollar headache for many cities that face a multiplicity of other problems, from population migration and the loss of tax base, to traffic, crime, and schools. The cities simply cannot afford the additional cost of replacing the urban infrastructure of sewers, despite litigation by the U.S. Environmental Protection Agency to force resolution through the federal court system.

The application of LID concepts to these communities is driven by the simple fact that if rainfall is prevented from discharging to the combined sewer, it

Figure 3-15 Sewer construction in the late 1870s buried most urban streams and turned them into sanitary and storm sewers, with combined discharge to tributary rivers.

Table 3-2 Portland CSO Program Costs, 1993–2010

Cornerstone projects	145 million
Columbia Sough	160 million
West Side CSO program	410 million
East Side CSO program	640 million
Total	**$1.4 billion**

Note: Construction of the Cornerstone projects has included infiltration sumps (3,000), downspout disconnections (50,000), local sewer separation efforts, and two stream diversions. The total annual reduction in stormwater thus far is 2 billion gallons.

will not overflow. The idea of keeping the rain out of the sewer is simple but has not been given serious consideration until the past 15 years, and only in a limited number of communities. One community that has led the way in this effort is the city of Portland, which under the leadership of the Department of Environmental Services [3] began a comprehensive effort in the early 1990s to both solve the problem of "conveyance" and to include program elements that would reduce the inflow of rainfall into the system. Over the past two decades, the city has invested over $1.4 billion in this program, most of which has been spent on the "plumbing" (Table 3-2) but 10% of which has gone into the "green" program elements. Of these investments, the least expensive and most successful has been the disconnection of rooftop downspouts (Figure 3-16), as well as the construction of two-manhole sewer inlets that infiltrate runoff, described as *sinks* (Figure 3-17). These two measures have combined to prevent some 2 billion gallons of runoff from entering the city sewers each year. More important, the city's program after 2012 will be comprised entirely of such measures, including an aggressive effort to infiltrate rainfall in vegetated designs (Figure 3-18), to accommodate not only new development but also existing communities where full stormwater mitigation has not yet been accomplished.

Urban streets that reduce or prevent rainfall from entering the sewer system have been described as "green streets" (Figure 3-19), the most conspicuous of which are streetscapes blended with planting beds or tree clusters, underlain by infiltration beds. Other urban LID designs utilize a combination of LID technologies, from vegetated roof systems to enlarged downspouts to containment and storage tanks to porous pavements, sidewalks, and walkways, all of which are underlain by aggregate or other media to hold, infiltrate, and slowly release rainfall from urban surfaces.

The reestablishment of surface vegetation comprised largely of trees is a major element of all green streets efforts. Here the goal is not only to restore a significant amount of pervious surface within the urban landscape, but also to create a tree canopy that provides environmental benefits well beyond the infiltration of rainfall, The direct cooling effect of a natural tree canopy, suggested to be on the order of 40% at full development, is dramatic. Of course, the actual land surface required to support this amount of canopy is more on the order of 15% of the land

Figure 3-16 The Portland "downspout disconnect" program has made 50,000 such changes and prevented over 1 billion gallons per year from entering their combined sewers. (Courtesy of Portland Department of Environmental Services).

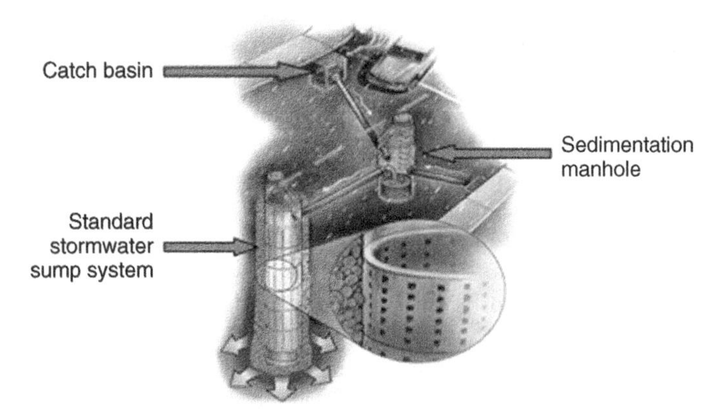

Figure 3-17 Infiltration sump design used in Portland, Oregon. (Courtesy of Portland Department of Environmental Services).

Figure 3-18 Vegetated infiltration bed in a Portland street. (Courtesy of Portland Department of Environmental Services).

surface, but should reduce the ambient temperature at street level significantly by providing constant evapotranspiration cooling during the warmest months of the year.

The habitat and aesthetic benefits of tree restoration in urban environments are more obvious, but in terms of water balance, this type of street design will allow the rainfall to soak into the soil mantle while the vegetation also returns it to the atmosphere by evapotranspiration. The subsurface design of a typical green street system must include a subsurface layer for rainfall storage and subsequent infiltration, with an overflow feature to assure that the bed does not become saturated and kill the root system. At the same time, because we are planting trees with deep root systems, we must take care to avoid any threat to existing structures, especially those with basements well below surface grade. Therefore, the creation of green street vegetation must be done on a site-specific basis, carefully fitted to the existing or reconstructed urban landscape.

While some forward-thinking cities such as Portland began developing green street programs two decades or more ago, a number of other communities have also initiated green streets and tree planting efforts, and in later sections we describe in detail how some have approached the solution. They all share one concept, however: The primary design goal is the reduction of runoff volume from the urban landscape, by infiltration, evapotranspiration, and/or capture and reuse.

End of the Sewer

Building beyond the urban fringe into the undeveloped countryside means that we must create new infrastructure systems of water, sewer, and drainage, not even

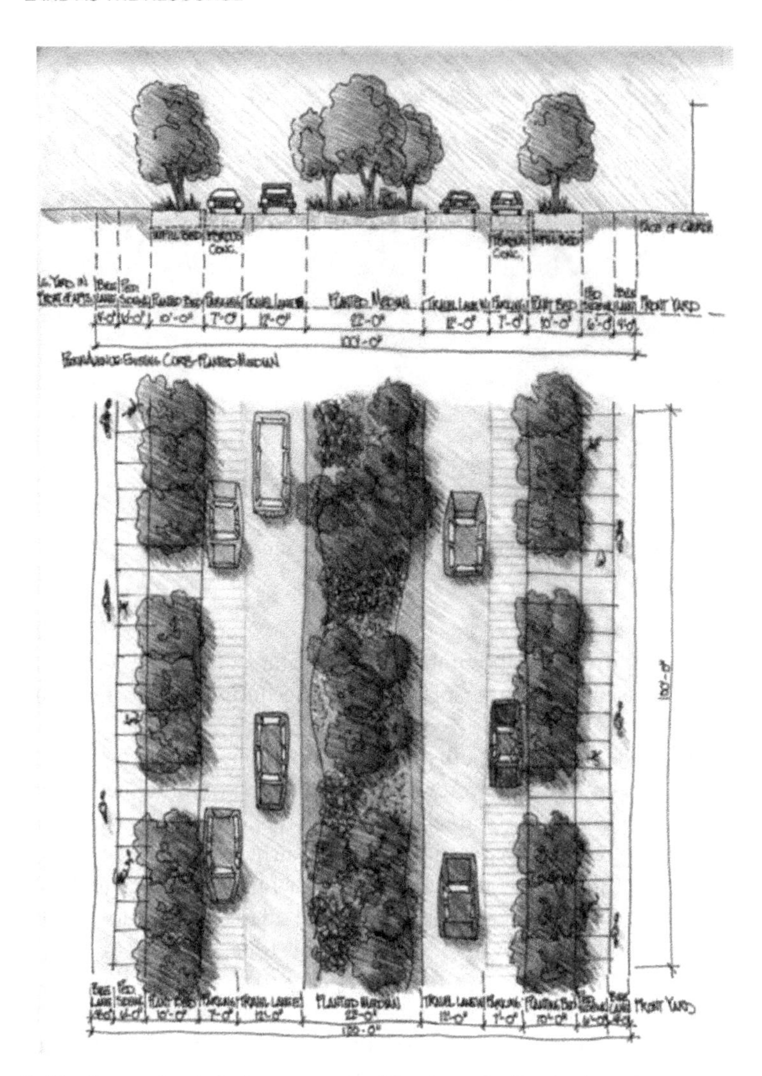

Figure 3-19 Green street design concepts utilize tree planting beds and porous pavements in various combinations within the right-of-way, depending on the available geometry.

considering the infrastructure of transportation, energy, and communication. This "urban sprawl" that we all criticize is in fact a reflection of our own perceived needs and represents the response of the marketplace to that demand. The pattern of land use resulting from these perceived demands is one in which growth frequently has followed the sewer rather than the opposite pattern. The building of sewers in many relatively undeveloped communities has been a guarantee of future growth and greater density of units, as the constraint of on-site sewage treatment is overcome.

During the 1960s and 1970s, the push to regionalize our wastewater systems and create a more efficient treatment process led to vast and complex conveyance systems that carry our sewage miles and hours before they reach a plant location, where they are then passed through a series of unit operations to reduce the pollutant load. Although there is certainly an efficiency of scale with larger treatment plants, where the treatment process can be closely managed, the hard question asked today is: What should be the limit of such regionalization designs? Would we be better served by wastewater treatment systems that confine the water supply and wastewater treatment process to some hydrologic scale (the watershed), within which the water used was returned to the local drainage rather than being exported to a larger watershed, well removed from any benefit of effluent reuse?

As we rethink our use of water as a finite resource, it is obvious that the scale of the design solution will have a great impact on our ability to achieve any type of restored natural balance. The optimal design, of course, could be to convey our sewage no further than is practical in terms of infrastructure cost and efficiency. Future designs may require that we no longer simply discharge the treated effluent to the local stream or river, but actually pump it back to open lands within the drainage, where we could infiltrate the water to the soil mantle, or use it for landscape irrigation, creating a kind of closed loop. In arid environments such as southern California, this type of effluent recharge is in fact already taking place and is proving to offer a partial solution to the resource limitations, but the energy cost of pumping effluent to upgradient locations must be given careful consideration. Again, the water–energy issues cannot be treated separately.

Is this effluent reuse too extreme for our modern society to embrace as part of a land development process? In the physiographic regions and arid environments where water is limited and demands exceed the regional and natural capacity, the application of such concepts seem to offer the only way to sustain the current demand for water without investing too much of our limited energy supply in more consumptive processes such as desalinization. One way of thinking about these types of designs is that they are based on sustaining the hydrologic cycle, even in those regions where the capacity of that cycle is limited.

Other Urban Infrastructure

For many sites, the watershed address is an important condition of new development and defines the water resource opportunities. First and foremost, does a safe and secure water supply source exist to meet the anticipated need, or could the extension of existing utility systems meet the demand? Does the local service provider utilize raw water sources from within the watershed, or is water imported from another watershed; and is this source adequate for the future? The great advantage of redevelopment in urban areas is the fact that the plumbing is in place and presumably is sufficient to meet the demand.

In developing communities where a sewer infrastructure exists, the sanitary sewers define the pattern of future growth, in a dense and continuous development pattern. On the other hand, an expanding road network allows discontinuous growth, as we all try to escape the urban center. The satellite communities are, however, insufficient as simple residential places and require the full complement of other uses. Everybody wishes to live in "the countryside," without any loss of convenience, so we bring our employment and marketplace to the new neighborhood, leaving behind the aging city and its compact conveniences.

Perhaps the best (or worst) example of how a growing region is comprised of multiple communities that evolved from an urban center is Los Angeles, now comprised of numerous residential neighborhoods and employment centers. This sprawling maze of politically separate places, each with a somewhat different character and quality, is bound together by a complex and intimidating roadway system that exceeds capacity on a regular basis, as residents try to move from home to work or retailing. This random and conflicting pattern of land use raises the question: Have we exceeded the limits of sane development at this place on the planet? The water supply, wastewater, and stormwater infrastructure that has evolved follows the same chaotic pattern, with numerous providers, both public and private, utilizing water resources from distant watersheds, collecting and treating the wastewaters in dozens of separate systems and largely ignoring the infrequent consequences of rainfall runoff to coastal waters, except when the soil adds the element of mud slides and affects residential communities.

3.8 THE LIVING BUILDING AND ZERO NET WATER USE

If LID provides the framework for sustainable water resources management, we might ask if it is possible to develop an even more sustainable concept that is based on a *zero net water use* principle. At first thought, this would seem to be a contradiction of the land development process; the old adage, "you can't make an omelet without breaking some eggs" comes to mind. Is it technically possible to build on a parcel of land without any net increase in water use beyond the natural hydrologic cycle? While such solutions may be possible for new development in a watershed without an existing infrastructure, the design of net zero water use in an urban environment would seem to be close to impossible.

The Cascadia Branch of the U.S. Green Building Council recognized the need for taking this next step in building technology in the late 1990s and proposed what they called the *Living Building Challenge* [4]. The concept was simple: Design buildings (and sites) that meet all of the required needs of the structure but create no additional demand for water and energy.

Over the past decade, some 60 building designs, largely in the northwestern United States, have attempted to meet this challenge. The concept has now begun to find broader application throughout the country, but few designs have had success. The zero net energy designs have developed many innovative concepts, focusing primarily on alternative energy sources (e.g., photovoltaic, geothermal, wind), and energy-efficient building (HVAC) systems or building materials. A great deal of valuable information can be found on the Cascadia Web site, or from their staff, consultants, or design professionals. In practice, the energy source can be situated remotely or on site, as long as it does not use fossil fuels.

The zero net water use is of primary interest here, in that it must go a step beyond the concepts of LID presented in this book and consider the impact and use of water through the entire hydrologic cycle. The challenge is to design a building that can meet the daily demand for fresh, potable water, recycle the resulting effluent within the building design for nonpotable uses or recharge the soil mantle, and capture the rainfall as the primary water source if groundwater sources are inadequate. If an external water supply is used, the issue is how to return this supply to the natural environment rather than to a sewer.

Applying the concept of zero net water use to new site designs begins with the question of whether or not public water (and sewer) is available. This is a prime determinant for most new land development projects and greatly simplifies the regulatory aspects of land development, not to mention the capital investment in infrastructure. Following the Living Building Challenge, however, the availability of municipal water and sewer services would not be a key consideration in the site selection process, and actually might present difficulties.

In an existing urban or suburban environment, the redevelopment of a parcel and new structure that meets the Living Building Challenge becomes complex, since the water and sewer service is long established, and any disconnection from this plumbing might present financial and regulatory problems for the developer. The public investment in existing infrastructure requires that any new development help to underwrite the operating cost of such systems, and if new development is allowed to "opt out" of the public systems, the long-term implications of new construction could be negative for the community as a whole. This is especially true for older cities that have gone through the cycle of decline and rebirth, which would hope to encourage redevelopment by the availability of such infrastructure.

In Pittsburgh, Pennsylvania, an existing complex of buildings, situated on a central hilltop, has been a major cultural asset to the community for over a century. Known as the Phipps Conservatory (Figure 3-20), this Victorian arboretum occupies over 13 acres, with the original glass conservatory buildings complemented by recent (2005) construction of a tropical rainforest building and large greenhouses for growing vegetation, as well as a meeting room and food service area. In 2008, Phipps decided to add an additional structure to the complex in a rear location that was in city ownership and used largely for vehicle service and materials storage. This building was to function as the Center for Sustainable Landscapes, and the goal was set to made it a living building, the first in Pennsylvania and possibly in the eastern United States. It was hoped that this center would develop and teach the cultivation methods and plant materials for sustainable landscapes, with little or no use of irrigation or of chemical pesticides or herbicides that would serve urban environments into the twenty-first century.

The goal of meeting the Living Building Challenge influenced every aspect of the design, and for over two years a team of architects, engineers, planners, and

Figure 3-20 The Victorian-era Phipps conservatory has designed a living building to serve as the Center for Sustainable Landscapes.

environmental experts from both the private market and local academic institutions worked to formulate a concept plan and detailed design. This effort led to a building program that was both unique and innovative for both the city and the region, and provided an important test of the living building concepts in an existing urban environment. One critical issue was the fact that the existing sewer system servicing the site was part of a century-old combined sewer, with major water quality impacts downstream on the Monongahela River from CSO discharges during wet weather. The city is served by a regional sewer system and plant downriver on the Ohio, but the original sewers were build in streambeds with stormwater inlets draining the surface and experience severe overflow and discharge conditions. The new living building would be served by a wetland treatment system, with an infiltration bed to avoid effluent discharge.

The concept plan included the capture of rooftop rainfall from the new glass greenhouses (Figure 3-21) to meet the major consumptive use of potable water in the complex for the irrigation of plants as well as the operation of display fountains. All new parking areas would be built with porous AC pavement underlain by stone beds for infiltration, with no runoff from the new site. The net result would be a significant reduction in the flows into the existing combined system, especially during wet weather.

This concept design, however, was not approved by city or state regulators. The urban soil in the building location included residual materials and was of poor quality for effluent renovation. Since the discharge of any effluent, no matter how small, was not acceptable under the living building criteria, the two regulations were in conflict. In addition, the proposed wetland treatment system of 400 gallons per day (gal/day) could not provide capacity for the much larger wastewater flow (3,500 gal/day) from the full complex, so the existing discharge

Figure 3-21 Runoff from the glass roofs will be collected in underground tanks for use as irrigation water.

to the combined sewer would continue. Although the issues remain unresolved and will probably lead to major program revisions, it serves to demonstrate how the dream of water resources sustainability clashes with the reality of existing urban systems, both physical and political. Change will inevitably come in all of the major cities burdened with CSO discharges, but not without similar conflicts for both water and energy resources.

REFERENCES

1. Perkins Eastman Consultants for the Pittsburgh Urban Redevelopment Authority, Pittsburgh, PA, 2011. *Allegheny Riverfront Vision Plan*.
2. Benke, A. C., and C. E. Cushings, Eds., 2005. *Rivers of North America*. Elsevier, Burlington, MA.
3. City of Portland, 2002. *Stormwater Manual*. Bureau of Environmental Services, Portland, OR.
4. Cascadia, 2009. *The Living Building Challenge*. Cascadia Chapter, U.S. Green Building Council, Seattle, WA. http://www.cascadiagbc.org.

4

THE PLANNING PROCESS FOR LID

4.1 SUSTAINABLE SITE PLANNING PROCESS WITH STORMWATER MANAGEMENT

Whether a site is currently undeveloped or urban, the sustainable site planning process must follow a few simple guidelines to understand both the limitations and the opportunities presented by a given parcel of land.

- *Guideline 1*: Understand the site and its watershed setting/context.
- *Guideline 2*: Apply conservation design: fit the program to the site.
- *Guideline 3*: Manage rainfall where it originates; do not convey.
- *Guideline 4*: Design with construction and maintenance in mind.
- *Guideline 5*: Calculate runoff volume and water quality.

Guideline 1: Understand the Site

Very few, if any, new development projects begin by asking: In what watershed is this parcel located? It simply is not part of the current strategy of the development community unless it is driven by some zoning or government regulation that recognizes a specific river basin or watershed as a special place, warranting careful land development controls. With millions of acres developed during the past four decades across the United States, few, if any, property owners and

Low Impact Development and Sustainable Stormwater Management, First Edition. Thomas H. Cahill.

developers have considered if a proposed plan can actually be sustained by the land that it will occupy without depleting the local hydrology. It is taken for granted that government will convey water to the parcel, carry away the sewage, and not be concerned if stormwater runs from the new landscape without any real restraints or sense of loss.

One could develop a list of questions to be answered for each development proposed:

- What watershed or hydrologic address identifies my parcel?
- What are the availability and limitations of this hydrology?
- Will my development program remove or destroy the vegetation?
- Will my development program require major earthwork and compaction?
- Will I utilize a surface source from a local river, or import water?
- Will I extract groundwater, and can I return it to the local aquifer?
- Will I need to convey wastewaters outside the local drainage?
- Can I clean the wastewater and reuse it by recharge design?
- Will my program produce increased runoff volume?
- Will my program generate surface pollutants in the runoff?

This set of questions is concerned with the water management issues and does not address the energy limits of a given site location. A similar set of initial questions could be developed that raise the same basic issues concerning energy, how it will be provided, and what limits and opportunities exist at a given parcel.

Guideline 2: Apply LID Conservation Design

As with any problem, it is usually easier to prevent it rather than to solve it. Many of our standard development practices create a given resource impact, and this impact can be largely reduced or eliminated by what we might call *conservation design*. In its most simple explanation, this translates as "fit the program to the site" without destroying the land and water resources that make this place suitable for development.

If the development program is residential in form, configure the unit layout to use as little space as necessary to achieve a community that satisfies the market conditions proposed: Cluster lots, minimize setback dimensions, reduce structural footprints, use only that portion of the parcel necessary to meet the design, and "drape" the plan over the existing landform, minimizing the need to reshape the land. Many developers take a very different approach to the land development process, clearing all existing vegetation and reshaping the land to meet a preconceived notion of what the final development program should look like from a given perspective (usually, a roadway entrance or highway). This type of land development process is frequently described as "terra-forming," with no small amount of sarcasm implied. The reference is to a science fiction theme that removes all existing natural features from a given planet planned for human occupation and forms a desired habitat prior to settlement.

In a situation where the terrain varies significantly over a parcel, woodlands and steep slopes, especially the combination, should be avoided for development. Designs that attempt to terrace such locations usually destroy the natural land system to the degree that soils erode rapidly and severely following regrading, despite the application of our best erosion control measures.

Guideline 3: Manage Rainfall Where It Originates

This concept turns the standard site design approach on its head. For decades, we have designed roof drainage systems, roadways, and grading surrounding buildings to remove and drain rainfall away from all structures as quickly and efficiently as possible. In fact, this was the beginning of the stormwater problem for both quality and quantity, because the immediate runoff of rainfall from impervious surfaces is not only the underlying cause of downstream flooding but also the primary transport mechanism for pollutant transport.

Although we cannot allow rainfall to infiltrate back into the soil directly beneath a building foundation, we can utilize the built surfaces that generally surround the structures, including all pavements, such as driveways, parking lots, local roadways, and the lands adjacent to these pavements that will be revegetated and landscaped. The commercial "big box" is the best example of this opportunity, since the pavement created for customer parking is usually much larger than the building footprint required (Figure 4-1) and offers ample opportunity for infiltration design with porous pavements, as long as the grading does not compact the soil mantle. One important consideration with the one-story retailing center is the perceived notion that every store must be approached from a flat pavement surface so the customers can move swiftly and easily from store to car with a cart full of purchases. The end result is the desire for as flat a site as possible, regardless of the natural landform.

Residential development is a more difficult setting in which to situate stormwater infiltration opportunities, since the associated pavements are smaller, with the largest pavement surfaces comprised of shared driveways and roadways (Figure 4-2). Landscaping elements that infiltrate rainfall, especially rooftop runoff, find greater opportunity in residential settings (Figure 4-3) and have been given the appealing label *rain gardens*. Some designs have used common site elements, such as cul-de-sacs (Figure 4-4), to infiltrate rainfall. A number of efforts have also been made to build detention basins that also infiltrate, but most of these have proven unsuccessful, with soil compaction and high water-table constraints limiting the long-term infiltration capacity and accumulating sediment fines clogging the basin bottom.

Stormwater conveyance must also be rethought. The conventional storm sewer system that carries runoff from rooftop and roadway in ever-larger pipes to the nearest surface water body is no longer a sustainable solution. A storm sewer system of perforated pipes that leak like sieves produces a very different end

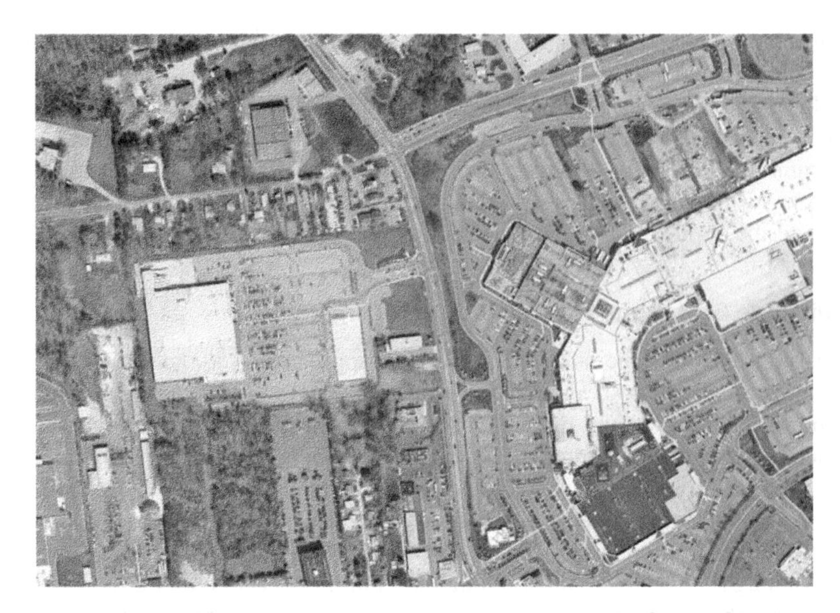

Figure 4-1 Traditional shopping center site plan, Anne Arundel County, Maryland.

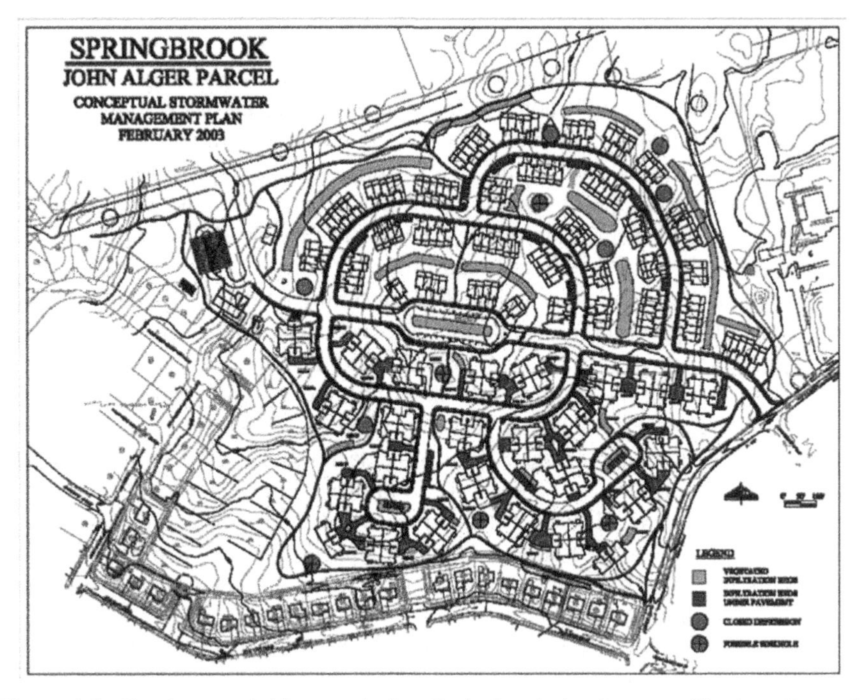

Figure 4-2 Roadway and driveway design, Springbrook development, Hershey, Pennsylvania.

Figure 4-3 Residential rain garden.

Figure 4-4 Cul-de-sac infiltration system.

result. The conventional system gets all of the stormwater runoff to the end of the sewer piping as quickly as possible, while it is the intent of the conservation design system to reduce almost all of the runoff to zero and have the storm sewers function as linear infiltration trenches (Figure 4-5).

Guideline 4: Design with Operation and Maintenance in Mind

Every engineering design begins with a basic method that is applied to solve a given problem, with new (or rediscovered) materials developed to implement

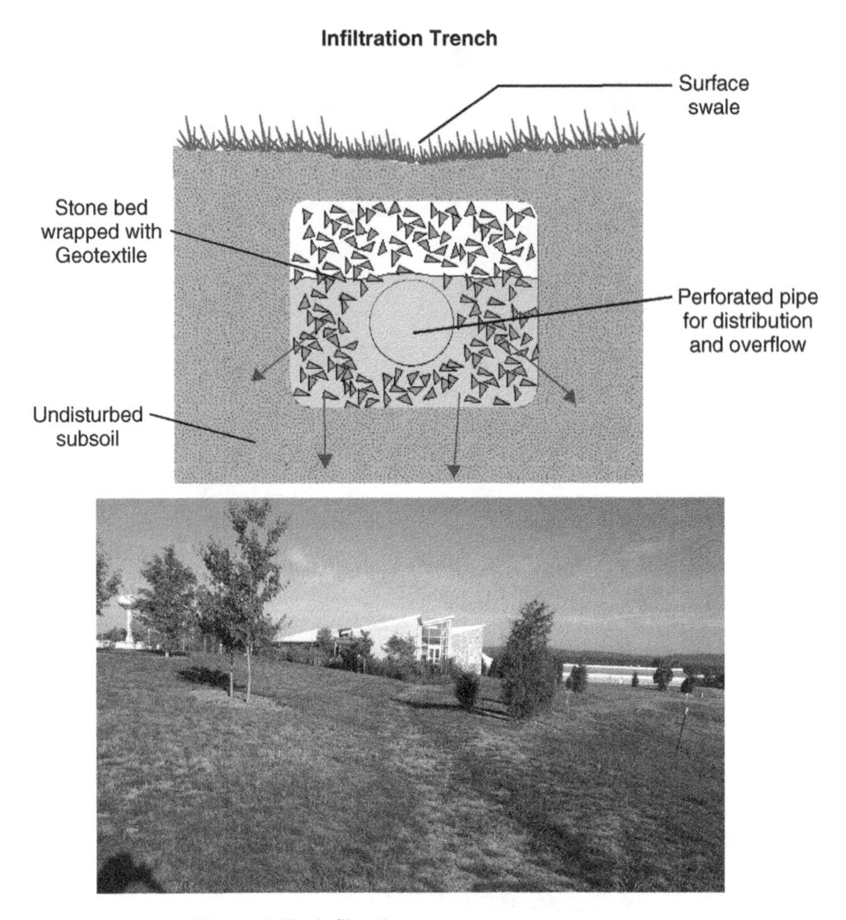

Figure 4-5 Infiltration system as a storm sewer.

this method. "Methods and materials" are the basic elements of every innovative solution. The actual implementation of this solution, however, must make a successful transition from concept to construction, using materials that have a proven capacity to function as designed and be capable of continuing to operate for the economic life of the project. The materials used must also be available in the local market and, if possible, be the product of a local economy.

The primary method advocated in this book as a solution for stormwater management is the volume reduction of stormwater, as compared to the current method of runoff detention. The most efficient way to accomplish volume reduction is infiltration of runoff into the soil mantle, a method that has been under development for over 35 years and mimics the natural processes that have shaped the planet for millennia. Many of the materials described here have proven to be efficient over years of actual construction experience in hundreds of projects over the past 30 years (Figure 4-6), while others are still evolving.

Figure 4-6 Porous pavement and infiltration system, Morris Arboretum, 1984.

As with all human-made materials, maintenance is a constant concern. Every design must consider this element in developing a solution, and with stormwater management it is a critical issue. Any method that removes sediment and other pollutants from a turbid fluid flow like runoff must, of necessity, include the maintenance of such systems in the implementation process.

Every design solution must also be practical and realistic, capable of fitting into a wide variety of site settings. Selection of the method is the first step in sustainable site design, but it must fit the site conditions. It must be a practical method to apply, and local construction personnel must be capable of building the design, following a carefully crafted specification.

The availability of required materials in the site location is important unless other suitable materials can be substituted. In the following chapters we propose methods that are universal in nature and materials that have been available at a given site, but the creative design process should never be ignored, and each designer must make use of his or her own skills to shape a solution.

Guideline 5: Calculate Runoff Volume Increase and Water Quality Impacts

A concept design proposed for a given site is first evaluated to determine the potential increase in runoff volume during a series of rainfall events, and then that site plan is evaluated to determine if dimensions, layout, and development needs can be modified to reduce the footprint of impervious surfaces. Other "nonstructural" measures are also considered for the building program. Then the modified site plan is evaluated to consider the type of structural materials that are suitable for site conditions, based on analysis of soils, geology, landform, and vegetation. Next, the potential pollutant production from the built site is estimated, and the various structural measures are evaluated to determine if they will be completely effective in preventing any increase in non-point source load generated from the parcel. Finally, the site plan is evaluated in the larger context of the watershed: whether it will have implications for other water resource impacts, such as water supply and wastewater, or perhaps produce secondary impacts, such as major roadway requirements or infrastructure expansion. The details of these steps are developed below, but the basic strategy should remain the same: Avoid, reduce, and mitigate stormwater.

In arid watersheds such as the southwestern United States, the value of rainfall for capture and potential reuse is too great to ignore, so that the method becomes the first consideration, especially with respect to rooftop runoff, which is an order of magnitude cleaner than street and gutter drainage. The prioritization of LID methods will always change to reflect the local hydrologic cycle. Many Southwestern communities, large and small, are severely limited in terms of water resources, and their approach to stormwater management should begin with a simple concept: "Capture the rain." This is especially true in regions such as southern California, where the population growth and land development have vastly exceeded the water resources available.

4.2 OVERVIEW OF THE SITE DESIGN PROCESS FOR LID

The site design process for LID builds on the site planning process, which occurs on the local level in most municipalities. The site design process for LID can be thought of as 10 basic steps that are illustrated in Figure 4-7 and summarized in Table 4-1. The 10 steps describe the entire development process from initial consideration of regulatory conditions through final construction and maintenance. Although the specific design solutions will vary, the process should be the same under most circumstances. Each step could be the subject of lengthy discussion, and articles have been written on many aspects of this process. However, to lead the designer through the site design process, the checklist format shown in Table 4-2 serves the need well.

An essential objective of the LID site design process is to maximize stormwater runoff "prevention" through the use of nonstructural best management practices (BMPs). When these measures are applied to the initial site design, the remaining volume reduction need can be met by structural BMPs, for volume control, water quality protection, and peak rate mitigation. An array of structural LID measures are set forth in Chapter 6, including porous pavements with infiltration beds, vegetated surfaces with infiltration beds, rain gardens, infiltration trenches, constructed wetlands, vegetated roofs, rain barrels and cisterns, and others.

The site design process for LID is intended to promote development of optimal stormwater management solutions in a cost-effective manner. Applicability of some parts of the process can be expected to vary. For example, some nonstructural BMPs can be expected to be challenging to apply in those cases where higher densities or intensities are proposed on the smallest of sites in already

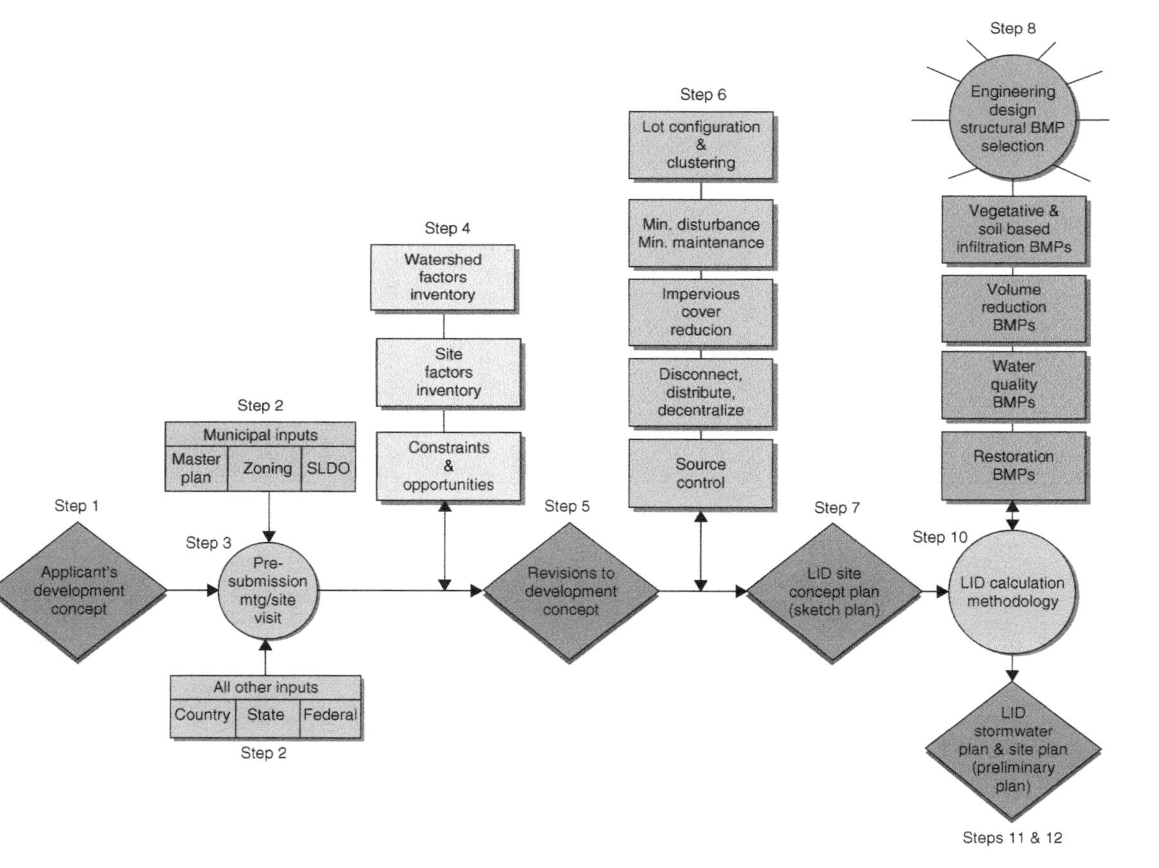

Figure 4-7 Site design process for LID.

Table 4-1 Site Design Process for LID

Step 1. A developer/builder/applicant decides to develop.

Step 2. Municipal, county, state, and federal requirements are incorporated into the development concept.

Step 3. A presubmission meeting is held as well as a site visit between applicant and local government representatives.

Step 4. The applicant inventories and evaluates the site and its natural systems (including the larger watershed context), highlighting constraints, natural system opportunities, and so on.

Step 5. The development concept is reviewed and revised, as necessary.

Step 6. Nonstructural LID measures are applied to the revised design concept.

Step 7. The applicant develops an integrated LID site plan concept or sketch plan.

Step 8. The engineering design/structural LID selection process.

Step 9. The stormwater calculations methodology (described in Chapter 6 and Appendix A) is completed so that structural measures achieve the required standards.

Step 10. An LID stormwater plan and LID site plan are developed.

developed areas, and where more highly engineered structural LID measures (Figure 4-8) may have to dominate the comprehensive stormwater management plan for LID.

Each municipality may want to adjust the site design process for LID to fit its particular needs. Perhaps the most important issue is the realization that the LID process involves *total* site design. Conventional stormwater management has often been relegated to the final stages of the site design and overall land development process, after most other building program decisions have been made Thus, stormwater management practices are frequently relegated to what appears to be the "leftover" areas of a site. On the contrary, the site design process for LID pushes stormwater management to the front of the line—into the initial stages of site planning process, when the development plan is being fitted and tested on the site. In this way, LIDs vital objectives and a truly comprehensive approach to stormwater management can be integrated effectively into the site design process.

The site design process for LID can be thought of as a series of questions, structured to facilitate and guide an assessment of site natural features, together with stormwater management needs (nonstructural and structural) of various land development concepts. The site design checklist for LID is intended to help implement the site design process and provide guidance to the land development applicant, property owner, or builder/developer in terms of the analytical process that needs to be performed as the development proceeds—the questions that should be asked and the answers that are needed—in order to formulate a truly low-impact development concept for the site. An especially important part of the site design process is the "testing" and "fitting" of preventive nonstructural

Table 4-2 LID Site Design Process Checklist

Step 1. Applicant's Concept Development

Step 2. Regulatory Guidance

Municipal Inputs: Master Plan/Zoning/SLDO

- Consistent with guidance, requirements, and options?
  - What does the comprehensive plan say?
  - What does the ordinance say?
  - How much of what goes where?
  - Existing zoning district?
    - Total number of units
    - Type of units
    - Density of units
    - Any allowable options?
  - Areawide clustering/concentrating?
  - Are resource systems protected?
  - Should the master plan or zoning be saying something else?
  - Are LID solutions required?
  - Incentivized?
  - Enabled?
  - Prohibited?
- Consistent with subdivision/land development ordinance (SLDO) requirements and options?
  - Performance standards for neotraditional, village, hamlet?
  - Reduced building setbacks?
  - Curbs required?
  - Street width, parking, other impervious requirements?
  - Cut requirements?
  - Grading requirements?
  - Landscaping?
  - Erosion and sediment control?
- Should SLDO be saying something else in terms of indirect, though critical, stormwater elements?
  - Township SLDO/stormwater requirements?
  - Peak rate and design storms?
  - Total runoff volume?
  - Water quality provisions?
  - Methodological requirements?
  - Maintenance requirements?
- Should the SLDO ordinance be saying something else in terms of direct stormwater management?
- Does SLDO require a building program to fit the constraints or opportunities of the site's natural features?

(continued)

Table 4-2 (*Continued*)

- Is applicant submission complete or fully responsive to municipal zoning or SLDO requirements?
- Is useful interaction (meeting, site visit, etc.) occurring at the sketch plan or even presketch plan phases?

All Other Inputs: County, State, Federal

Step 3. Presubmission Meeting and Site Visit

Step 4. Inventory for Opportunities and Constraints

Watershed Factors Inventory

- Major/minor watershed/location?
- State stream use/standards designation/classification?
    - Special high-quality designations?
    - Required standards?
- Any Chapter 303d/impaired stream listing classifications?
- TMDLs?
    - Existing, planned?
- Existing/planned downstream sensitivities?
    - Water supply intakes?
    - Known downstream flooding?
- Known water quality trends in the major or minor watershed?
- Aquatic biota, other sampling, monitoring?

Site Factors Inventory

- Important natural site features?
- Existing hydrology (drainage, swales, intermittent)?
- Existing topography, contours?
- Existing subbasins?
- Soils, their hydrologic groups?
    - Seasonal high water table, depth to bedrock, hydric?
    - Alluvium?
- Geologic issues?
    - Carbonate?
    - Slide potential?
    - Earthquake zone?
- Existing vegetation?
    - Woodlands, forests?
    - Specimen trees, mature, other?
    - Shrub layer, meadows?
- Species of special importance?
    - Limited habitat?

Table 4-2 (*Continued*)

- Special value areas?
  - Wetlands?
  - Floodplains?
  - High-quality woodlands?
  - Riparian buffers?
  - Naturally vegetated swales, drainageways?
- Special sensitive areas?
  - Steep slopes?
  - Geologic constraints?
  - High water table?
  - Aesthetics?
  - Air quality, noise, other issues?
- Size and shape of the site?
  - Special constraints, opportunities?
  - Site border conditions?
  - Contiguous parcels?
  - Upslope drainage?
- Existing areas of known/suspected contamination?
- Existing development features of site?
  - Existing structures, if any?
  - Preservation, restoration, historical value?
  - Existing land cover?
  - Existing impervious areas?
  - Existing maintained areas?
  - Existing infrastructure (water, sewer, storm drainage)?
  - Other infrastructure (power, communication, energy)?

Constraints and Opportunities

- Constraint zones at site?
  - Watershed issues?
  - Special features?
  - Opportunities (well-drained soils, etc.)?

Step 5. Revision to Development Concept

Step 6. Nonstructural LID Measures

Lot Configuration and Clustering

- Reduced individual lot size?
- Concentrated/clustered units/lots?
- Lots developed to avoid or preserve sensitive features?
- Lots configured to avoid earthwork, fit topography?

(*continued*)

Table 4-2 (*Continued*)

Minimum Maintenance/Minimum Disturbance
- Defined disturbance zones?
- Protect sensitive areas?
- Sustain natural drainage?
- Minimize earthwork, cut and fill?
- Include reforestation, revegetation?
- Minimize soil compaction?

Impervious Coverage Reduced, Minimized
- Reduced road width?
- Reduced cul-de-sac diameter, open center?
- Reduced parking requirements?
- Reduced parking area size?
- Shared parking considered?
- Use of porous pavements, infiltration beds?
- Design single sidewalks, use porous pavements?

Disconnected, Distributed, Decentralized
- Disconnect roof downspouts from impervious surfaces?
- Utilize rain barrels, rain gardens?
- Design stormwater conveyance elements that infiltrate?

Source Control
- Street sweeping required?
- Lawn chemical application avoided, minimized?
- Irrigation only with captured runoff?

Step 7. LID Site Concept Plan (Sketch Plan)

Step 8. Engineering Design/Structural LID Measure Selection

Design Selection
- Runoff quantity needs?
- Runoff quality needs?
- Conveyance minimized?
- Impervious footprints minimized?
- Soil infiltration opportunities, capacity?
- Infiltration constraints?
- Construction cost comparison?
- Maintenance cost comparison?
- Ease of construction?
- Aesthetic benefit?
- Environmental benefit?
- Maintenance possible by tenant, owner?

Table 4-2 (*Continued*)

Step 9. Use Stormwater Calculations Methodology (Appendix A) to Design Structural Measures

Infiltration LID Measures

- Suitable soil permeability, depth, location?
- Location of structures, impermeable pavements?
- Placement of structures to maximize infiltration systems?
- Select LID measure based on site plan?
- Revise site plan to fit infiltration opportunities?

Other Volume-Reduction LID Measures

- Vegetated roof systems possible?
- Capture/reuse possible?
- Short- and long-term maintenance?

Water Quality Only LID Measures

- Conveyance structures (inlets, chambers, filters)?
- Vegetated detention basins?
- Constructed wetlands?

Other LID Measures

- Riparian restoration?
- Woodland restoration?

Achieved Principles of LID?

- Sustain the hydrologic cycle?
- Sustain existing trees?
- Sustain the soil mantle permeability?
- No net increase in runoff volume for 2-year rainfall?
- No increase in NPS pollutant discharge?
- Volume reduction with infiltration, evapotranspiration, and/or capture/reuse?
- Minimize impervious surfaces?
- Disconnect–decentralize–distribute?
- Use vegetative systems?
- Include maintenance?

Step 10. LID Stormwater Plan and Preliminary Site Plan

BMPs at the site. The checklist provides questions designed to facilitate the potential application of these nonstructural BMPs. Some municipalities may want to consider formalizing the checklist via ordinance so that applicants are required to submit a "checklist report" which demonstrates that the applicant—and most important, the proposed LID site plan and stormwater plan—has addressed these questions fully.

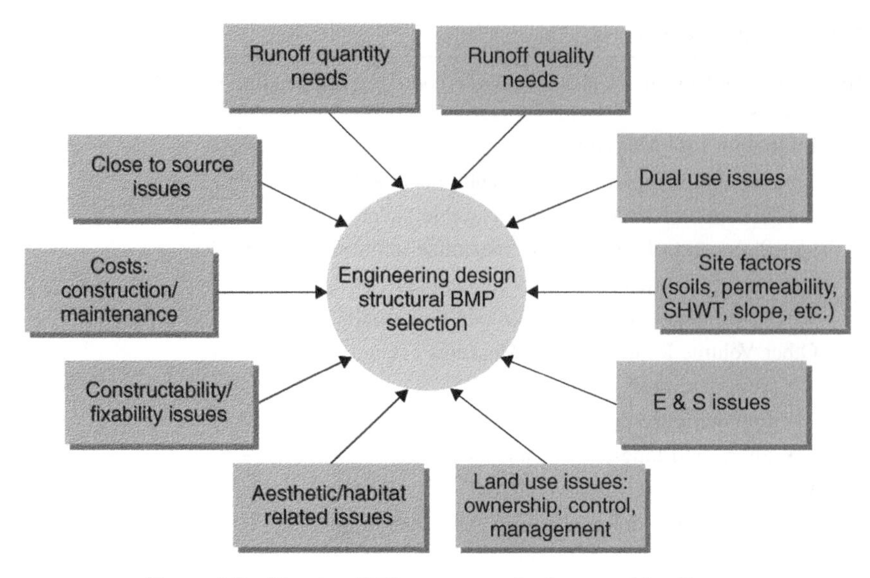

Figure 4-8 Structural LID measure selection considerations.

The goal here is to rethink the site design process to achieve low-impact development objectives. That's an ambitious goal. Design by its very nature is expansive and doesn't want to be compressed into a series of steps. Nevertheless, this proposed set of steps and list of checklist questions can provide guidance to move this important design process forward.

5

THE LEGAL BASIS FOR LID: REGULATORY STANDARDS AND LID DESIGN CRITERIA

5.1 THE LAND–WATER LEGAL PROCESS

It is intuitively obvious that what we do on the land has everything to do with the water resources on which we rely to sustain our environment and habitat. However, the system of laws, regulations established based on those laws, and design criteria evolved to comply with those regulations do not recognize the interdependency. It is the connection between land and water that has led to the formulation of such concepts as low-impact development, sustainable site design, LEED (Leadership in Environment and Energy Design) green buildings, "living buildings," and similar efforts to change or modify our land development process in order to reduce or mitigate the resulting environmental impact. This is true not only with respect to water but also to energy. The goal of LID and all similar programs is that by applying sustainable building concepts, we can provide a high-quality environment for future generations while creating places and spaces that meet all of our needs for habitat.

Common Law

When we consider the laws that regulate how we develop the land and use the rainfall, we begin with the common law. As described by Thomas Sullivan in his *Environmental Law Handbook* [1]:

Low Impact Development and Sustainable Stormwater Management, First Edition. Thomas H. Cahill.

> Underlying the development of legal theory in the United States is a body of rules and principles related to the government and security of persons and property which had its origin, development and formulation in England. Brought to the American colonies by peoples of Anglo-Saxon stock, these basic rules were formally adopted in the states in which they were in force after the American revolution. Known as the "common law," these principles are derived from the application of natural reason, an innate sense of justice and the dictates of conscience. The common law is not the result of legislative enactment. Rather, its authority is derived solely from usages and customs which have been recognized, affirmed and enforced by the courts through judicial decisions.
>
> It is important to realize that "common law" is not a fixed or absolute set of written rules in the same sense as statutory or legislatively enacted law. The unwritten principles of common law are flexible and adaptable to the changes which occur in a growing society. New institutions and public policies; modifications of usage and practice; changes in mores, trade, and commerce; inventions; and increasing knowledge, all generate new factual situations which require application and reinterpretation of the fundamental principles of common law by the courts.
>
> As the courts examine each new set of facts in the light of past precedent, an orderly development of common laws occurs through a slow and natural process. Thus the basic principles underlying American jurisprudence remain fundamentally constant, evolving slowly and progressively in scope as they absorb the surface ripples produced by the winds of social change.

The American legal system has evolved two very different bodies of law that define how the public (and private) interest is protected when land is developed as compared to when water resources are affected. The regulatory processes and the administrative systems that guide decisions concerning land use and water use are separate and independent of each other, with state and federal laws protecting water, and local units of government deciding exactly how the land is to be used. Of course, there are some overlapping areas, but for the most part, the decision makers for land use are different from those who control the use, allocation, and quality of our water resources. To illustrate, we consider how environmental law has evolved with respect to water quality and land use in the United States.

Federal Water Quality Law

Prior to 1970, the states set standards that established allowable concentrations of pollutant parameters for various water bodies. In 1972, the federal effort to restore water quality began with the passage of the Federal Water Pollution Control Act [2] and the Marine Protection, Research and Sanctuaries Act [3]. In 1977, the Federal Water Pollution Control Act was renamed the Clean Water Act (CWA), and Congress changed the regulatory focus to rigorous control of toxic water pollutants. In 1987, extensive amendments were made to assure attainment of water quality goals in areas where compliance was insufficient. From this foundation, a large set of regulations related to various permitting processes has become the primary guidance throughout the country, with the delegation of most programs to state agencies.

The primary goal of the CWA was to eliminate the discharge of pollutants to surface waters by the year 1985, a significant target date. This resulted in a program to provide permits for the discharge of wastewaters that were subject to processes that would eventually eliminate such contaminants from the wastewater effluent, both municipal and industrial. These were labeled National Permit Discharge Elimination System permits, and a lengthy process of standard setting was initiated, based on the pollutant removal technology then available. Of course, the goal of elimination of all pollutants was never achieved, nor is there any hope that it will be; however, the concept of permits to limit the quantity and quality of wastewater flows and to regulate the treatment processes required has become the foundation of all federal regulations intended to achieve a sustainable level of quality in the nation's waters.

Within the past decade, the recognition that stormwater runoff comprised a significant fraction of the pollutants that were entering surface waters resulted in modifications to regulations based on sections of the original CWA (Section 303.d) that required municipalities with multiple discharges from separate storm sewer systems (referred to as MS4) to obtain and maintain a permit for such discharges, including consideration of ways to reduce the pollutant load conveyed to surface waters. Foremost in these requirements was the use of LID technologies, although no specific load reduction was established for a given waterway or permit. The application of this MS4 permit criterion has been uneven across the United States, with coastal waters (southern California) and estuaries (Chesapeake Bay) receiving the initial regulatory attention. The LID technologies described in detail in the following chapters play a critical role in achieving compliance with this federal water quality criteria.

Federal Land Use Law

The federal government has carefully avoided any legislation that would restrict state land use controls. However, passage of the National Environmental Protection Act (NEPA) in 1969 [4] declared a national environmental policy that promoted consideration of environmental concerns by federal agencies, both on the substantial lands held by the federal government in the Western states and the numerous military bases and other federal properties and buildings situated in every state. In effect, any action that is funded or supported by federal dollars or permitted by federal regulations must be in compliance with NEPA. Most important, the requirement that an Environmental Impact Statement (EIS) be prepared for any such action has proven to be a powerful tool in many land use decisions. Many states have in fact passed their own version of NEPA and set requirements for EIS preparation that directly involve land use actions.

In the following chapters, the recommended technical solutions described under LID assume consistency with federal and related state land and water law, but this is not always accurate at present in every jurisdiction. For analysis, two very different states are considered here: Pennsylvania, one of the original 13 colonies, with some 2560 separate units of government, each of which operates as an independent regulator of the land within its boundaries, and California, the most populous state and one of the more recent (statehood in 1903), blessed with an abundance of natural resources and constraints, not the least of which is an arid environment on the southern coast, where a major fraction of the state population now resides.

5.2 THE EVOLUTION OF LAND DEVELOPMENT REGULATION

For 15,000 years we have built our habitats with the natural materials locally available, based on the level of technology that a given society or culture possessed. As we evolved from the use of rock shelters (Figure 5-1) and caves (Figure 5-2), we developed skills to shape and form the wood and stone and found that the metals extracted from the earth could greatly assist us in these

Figure 5-1 Oldest human habitat structure: rock shelter, 45,000 B.C.

Figure 5-2 Cave habitats (10,000+ B.C.) evolved to residences (A.D. 1500).

tasks, ultimately becoming not only the tool but also the building material. A few simple concepts of structure evolved over centuries into complex building systems that allowed the development of monumental temples and palaces that we still admire and honor as great works of design, thousands of years later. The methods and materials evolved as the need increased, but other than location, the land served only as a convenient palate on which to craft our structures.

As hamlets evolved into villages and then into towns and cities, the land was a convenient surface on which to shape and mold our building systems, and served no major function, except perhaps as earthwork barriers for protection to complement or reinforce structural walls or to support our roadways as our transportation systems evolved. The value of the land for cultivation, however, was recognized over 10,000 years ago, and communities were situated convenient to fertile soils or as market locations. In fact, during the agrarian history of our species, the location of cities was driven to protect these food production resources, and to collect and barter them for other goods and materials. The use of land was centered on optimization of productivity for most of our recent history.

The first land use law enacted in English law, which provides the foundation for U.S. law, was passed in 1562 by Queen Elizabeth I and dealt with both land use and water. The problem was the frequent flooding of the Thames River and the use of lands within the floodplain held by various barons and dukes. The solution was to build a barrier or berm along the river, not unlike our system of dikes along the major rivers throughout the United States. To give the crown the power to enter into and build the dike on private lands, the Queen passed the "Law of Sewers" (*sewer* is derived from a Saxon word for seawall), and thus the first environmental law was created to protect the public interest from water impacts, based on the restriction of land use.

Until very recently, the concept of a system of laws to protect our natural resources was unheard of, and all land and water was to be exploited by the powers that controlled them. In fact, the distinction between the people who worked the land and those who owned or controlled it formed the basis of many, if not most cultures, over the past 3,000 years. The concept of serf and master that defined this system of land use has vanished from modern society, although some might argue that the economic disparity has not.

In the evolving United States, that distinction was no longer preserved, except in the southern states until the Civil War and the end of slavery. The idea that every person could own (and use) land as he or she chose was firmly rooted in this young culture. The American frontier was a seemingly vast and rich land with abundant natural wealth, or so it seemed to the earliest European settlers, who arrived, primarily on the eastern Atlantic coast of North America, some 380 years ago. Migrating from a set of cultures that had used the land and water intensely for centuries, they were largely interested in the woodlands and the potential of clearing the land for cultivation. As the government changed from colony to independent union, the loosely united "states" looked to their immediate western borders and considered more opportunities for resource exploitation.

After an additional century of expansion and integration, the United States had formed an independent system of government that was unique in the history of civilization. This union fractured under the issues of slavery and states rights after the initial century, only to reunite for the great expansion westward. During this final phase of national unification, the land became the motivation for settlement, and vast portions of the new territories (taken from the native cultures by conquest and confinement) were given away as incentive for settlement. Property was now controlled by the new settlers, who vigorously defended their rights of ownership.

All existing and new citizens had the legal right to use their land as they saw fit, and the extraction of mineral resources, from coal to gold, became a part of this right. As the new American landscape was taking form, those who cultivated and those who used the land for grazing of livestock had different ideas about both land and water use law, and local regulations evolved in each state to serve their needs.

The corresponding water laws that were formulated by states west of the Mississippi were structured differently from the Eastern water regulations, with the "right" to extract groundwater treated in a fashion similar to that of minerals and petroleum. Although the initial legal structure of the colonies was patterned after the English legal system, the evolving American legal process now perceived land as a right of citizenship, giving wide latitude to the property owner as to use and alteration, including removal of woodlands, grading and regrading of the soil mantle (including removal and filling), extraction of subsurface deposits, and other actions that effectively destroy the natural system. The impacts of these actions on the land's water resources, both surface and subsurface, have come to be regulated only to the extent that the public welfare is affected. Until recently, this water impact has been measured by changes in water quality, assuming that the degradation of surface waters is harmful to public health, and is protected under the common law.

The American vision of conquering the wilderness, clearing the forest, tilling the field, and damming the mighty river seemed reasonable as long as the natural resources of land and water were unlimited and could be exploited without fear of loss. However, our perception in the twenty-first century indicates that there is indeed a limit to our use of natural resources, and we are slowly crafting a new set of guidelines to sustain what remains. The concept of LID is informed by this recognition.

5.3 THE REGULATORY FRAMEWORK

The system of laws that regulate the use of land and the protection and management of water resources in the United States can be described in one simple phrase; they are totally different and unrelated in form and structure. Although we might all agree that what we do on the land has everything to do with what happens in the water, both surface and ground, the legal system in the United

States has never made the connection. Even those land activities that clearly pollute the water have not been prohibited, only regulated by permit. The institutions and agencies that make decisions and issue such permits concerning land use are structured within state and local government, by and large, and no federal effort to make land use decisions has ever been politically acceptable. This does not apply to those land holdings within the ownership of the federal government, which are quite extensive in the Western states. The Bureau of Land Management, the National Parks Service, the Department of Energy, and the numerous military bases and parcels distributed throughout the country make the federal government the largest single property owner in the United States.

The form and structure of land use decision making varies with each state, from the original 13 colonies to the midwestern prairie states to the far western states, with a changing attitude toward the rights and powers of local governments as the country grew westward. It would not be inaccurate to say that we have 50 different systems of land use management, with certain similar patterns and attitudes toward land regulation grouped by region.

It is of interest to consider the subject of LID as it can be applied within the institutional framework of two states, very different in background, origin, and history. The first is Pennsylvania, one of the original colonies and the first capitol of the United States in Philadelphia, founded by Quakers who sought to build a new government that would allow them religious freedom not found in their native England. The second is the state of California, created over a century later and evolved from a largely Spanish culture and colony. It has long been said that everything that happens in the United States happens first in California, both good and bad. The system of land use that evolved here is different from the older Eastern states, out of necessity as well as the geography of politics, with two portions, north and south, forming their own individual places.

Pennsylvania Land Use Law

The original northeastern colonies closely followed the English pattern of forming counties and further subdivisions of townships and boroughs, with a few large cities that became their own county (Philadelphia) or part of a county (Allegheny) in the case of Pittsburgh, surrounded by some 81 other municipalities within the county. For example, in Pennsylvania, an original "commonwealth," the state is divided into 72 counties with some 2560 municipalities, each of which operates as an independent entity when it comes to deciding how the land is used by its citizens. Many townships, but certainly not all, have established zoning, a system that determines what land use goes where (based largely on the road network). No consistent rule governs the use of land at contiguous points between adjoining municipalities, producing what can only be described as a disjointed pattern of land use types within a given watershed, stream valley, or county.

While county governments in Pennsylvania play some role in land use planning, especially where a county planning commission has been established, they have little real power or control. They support and operate a number of related agencies and departments that are critical to the independent municipalities, who rely for such services as public health, social services, institutions, and some infrastructure facilities (bridges, etc.). The decisions concerning the use of land in each of the 2560 municipalities is made by a small group of elected officials, supported by a cadre of professional lawyers, engineers, and planners, who provide the professional guidance to assure compliance with all regulatory criteria. From a constitutional perspective, local government has only those powers that the state has delegated, but it is unlikely that any future state government will have the political will to make land use decisions, with the exception of the highway department with respect to rights-of-way. This difficult and somewhat feudal system of land use management will continue into the foreseeable future.

Pennsylvania Water Law

Water Quality The commonwealth of Pennsylvania has a relatively long history with respect to protecting water quality in surface waters, driven by an equally long degradation of streams, primarily as a result of mining, both deep and surface, which has polluted some 10,000 miles of stream with acid waste. The mining of coal was long considered the economic backbone of this state, and during the early part of the twentieth century it could be said accurately that "Coal was King," and little was done to prevent the impact of residual coal seam material, rich in sulfates that oxidized to become sulfuric acid, on the regional surface waters. Finally, in 1937, the state enacted the Clean Streams Law specifically to address this issue, but also to prod municipalities into the building of wastewater treatment facilities. Over time and with strong federal support after 1972, the state has evolved a fairly good set of water quality regulations intended to protect the abundant water resources with which it is blessed. However, it has not restored the original quality to thousands of miles of streams.

Stormwater In contrast to some other states, Pennsylvania does not have laws and associated regulations that manage stormwater on a state level. Stormwater management in Pennsylvania is largely a "bottom-up" system where municipalities shoulder the brunt of the management as well as the regulation. Where municipalities lack zoning ordinances as well as subdivision and land development regulations, there may be little or no formal stormwater regulation on the municipal level. With expanding programs, such as NPDES Phase II and PADEP's Post Construction Management Permits/Plans for sites disturbing more than 1 acre, as well as Act 167 Stormwater Management Plans for designated watersheds, some upper-level management intervention by county conservation districts and PADEP itself is occurring, regardless of the municipal stormwater management program.

Although these new and expanding NPDES Phase II and Act 167 programs provide some level of consistency and uniformity for stormwater management in Pennsylvania, there nevertheless remains a basic lack of guidance around which the municipalities in the state can structure their respective stormwater management programs, because of this lack of a state law or regulation for stormwater management. The recent (2005) *Best Management Practices Manual* [5] and the related model ordinance are intended to provide that guidance but have yet to find full acceptance by local government throughout the state.

Given the many different development contexts and physical site constraints in Pennsylvania, it is reasonable to ask how we can regulate the development process to mitigate or prevent the resulting stormwater impacts. What performance standards do we set for the design of a proposed development site, including water quality, peak rate control and flood prevention, infiltration and groundwater recharge, and stream conservation and protection? How many best management practices need to be included? How large do they need to be? What level of performance should be achieved? In the simplest of terms, what standards should guide stormwater management planning in Pennsylvania communities? And perhaps most important, how do we link the land development process with these measures?

Both land developers and public officials must grapple with this essential issue of land management standards as municipal management programs and regulations are put in place. Because of the variability of Pennsylvania communities, standards have taken the form of ranges of values in some cases, reflecting natural system variability, as well as the vast differences between the nearly pristine watersheds and highly urbanized and degraded drainage areas that exist in Pennsylvania. In all cases, municipalities are encouraged to strive to enact the most rigorous management programs possible, even as issues of economic feasibility are taken into account and kept in balance. Even where Act 167 watershed plans or river conservation plans have been developed and adopted, individual municipalities should look to their watershed neighbors and strive to integrate their individual municipal actions with the watershed system whole.

California Land Use Law

While the government of California took form a century after that of Pennsylvania, the concept of leaving land use decisions to local government units still held, and the state does not retain control over most land decisions, although many policies have indirect influence. In 2008, Richard Watson, a professional planner working on the draft of the LID manual for southern California, summarized the system of land use management in California:

> [T]he cornerstone of the California planning process is the general plan. According to Thomas Kent, in his text *The Urban General Plan* [6], a general plan is "the official statement of a municipal legislative body which sets forth its major policies concerning desirable future physical development." The general plan process is defined by Government Code Sections 65000–66037, which delegate most local land use decisions to individual cities and counties across the state. Each county and incorporated city is required to adopt "a comprehensive long term general plan for physical development."

General plans include development goals and policies and lay the foundation for land use decisions made by planning commissions, city councils, or boards of supervisors. General plans must contain text sections and maps or diagrams illustrating the general distribution of land uses, circulation systems, open space, environmental hazard areas, and other policy statements that can be illustrated. The Government Code specifies that general plans must contain seven mandatory elements or components: circulation, conservation, housing, land use, noise, open-space, and safety. Local governments may also voluntarily adopt other elements addressing topics of local interest. Cities and counties could adopt an optional water element in their general plans, but few have done so. Instead, water has most often been partially addressed in either the mandatory conservation element or in optional natural resources or public facilities elements. Water is frequently addressed only in terms of water supply and/or water conservation.

California law establishes zoning as a regulatory mechanism to implement general plans. Zoning is adopted by ordinances and must be consistent with general plans. Under a zoning ordinance, development is required to comply on a lot-by-lot basis with specific enforceable standards. Zoning ordinances specify categories of land use and associated standards such as minimum lot size, maximum building heights, and minimum building setbacks. Zoning ordinances can include overlay zones that provide additional standards for specified areas such as historic districts, wetlands, and other areas deemed to require extra protection.

Traditional zoning is often referred to as Euclidian zoning after the United States Supreme Court decision that affirmed the legality of zoning to separate land uses. Separation of uses became widespread as zoning gained popularity. LID is not specifically addressed in traditional zoning, but some of the standards included in specific zones can provide significant barriers to implementations of LID practices.

Currently, the planning profession and many communities are experimenting with a different approach to zoning that provides more flexibility regarding building areas within particular zones combined with more stringent regulation of design elements such as architecture, landscaping, and pedestrian-friendly circulation systems. This form of zoning can help to implement smart growth, as was made possible by the approval in 2004 of Assembly Bill 1268, which allowed the use of form-based codes in the state. Form-based codes provide the flexibility to address LID and, in fact, invite the inclusion of detailed LID design elements [7–9].

California Water Law

Watson also described how LID was related to water quality regulations in California;

> [T]he use of LID measures in California is driven by water quality regulations and promoted by the Ocean Protection Council. The State Water Board's formal adoption of its Strategic Plan Update 2008–2012 restates the Board's vision of "a sustainable California made possible by clean water and water availability for both human use and resource protection." The update contains a sustainability principle and value that states, "we commit to enhancing and encouraging sustainability

within the administration of Water Board programs and activities by promoting water management strategies such as low impact development, considering the impacts of climate change in our decision-making, and coordinating with governmental, non-profit, and private industry, and business partners to further strategies for sustainability."

Starting with the San Diego MS4 Permit (Order No. R9-2007-0001), adopted on January 24, 2007, municipal permits within the region have contained specific LID and hydro-modification requirements [10, 11]. The major emphasis of the LID requirements in municipal permits is on reduction of impervious area in order to facilitate infiltration and reduce urban runoff. LID requirements in MS4 permits are to apply to specified categories of new development and redevelopment projects. The Permittees are tasked with the responsibility of developing design and maintenance criteria and establishing minimum standards for the use of LID practices. They are also required to develop manuals or technical guidance for municipal employees and private sector practitioners involved with the implementation of LID practices.

5.4 STORMWATER MANAGEMENT REGULATIONS

In the current stormwater management regulations that have taken form across the United States (and continue to evolve), the critical question is: How can we use LID methods and materials to achieve compliance with the federal and state rules and effectively mitigate the impacts of land development?

Volume Control

Once the water balance has been estimated for a given site, the guidelines for sustaining this balance must be established. These technical guidelines evolve into specific design criteria that can be applied to a broad spectrum of stormwater management solutions, commonly referred to as best management practices (BMPs) or low-impact development measures. This book presents stormwater management principles and recommends site control guidelines for volume, water quality, and rate. These guidelines are proposed as the basis for municipal stormwater regulation, and offer guidance for municipalities desiring to improve their stormwater management programs. Some state laws and regulations manage stormwater directly at the state level, while some state-level management occurs through programs such as NPDES Phase II permitting, or the MS4 program, but whatever the regulatory format, the first step is to "set the bar" for our LID strategies, not unlike any other type of design.

Although the volume control guideline is quite specific concerning the volume of runoff to be controlled from a development site, it does not limit the methods by which this can be accomplished. The selection of an LID measure, or combination of measures, is left to the design process. In all instances, minimizing the volume increase from existing and future development is the goal.

The BMPs described in this book place emphasis on infiltration of precipitation as an important solution, but this is only one of the three basic methods that reduce the volume of runoff from land development. These three methods can be summarized as follows:

1. Infiltration
2. Capture and reuse
3. Vegetation systems that provide evapotranspiration

All of the stormwater management systems described in this book include one or more of these methods, depending on specific site conditions that constrain stormwater management opportunities.

Volume Control Criteria

Volume control is essential to mitigate the impacts of increased runoff. To do this, volume-reduction BMPs must meet the following criteria:

1. *Protect stream channel morphology*. The increased volume of runoff results in an increase in the frequency of bankfull or near-bankfull flow conditions in stream channels. The increased occurrence of high flow conditions in riparian sections has a detrimental effect on stream ecology, including stream channel impacts and overall stream morphology, with stream bank erosion greatly accelerated (Figure 5-3). As banks are eroded and undercut and as stream channels are gouged and straightened, meanders, pools, riffles, and other essential elements of habitat are lost or diminished. Research has demonstrated [12] that bankfull stream flow in a natural channel typically occurs between the 1- and 2-year-frequency storm event (often around the 1.5-year-frequency storm). Land development will cause bankfull stream flows to occur more often than the 1-year-frequency storm event, and in highly urbanized watersheds (over 25% impervious cover), bankfull flow can occur as frequently as every few months (Figure 5-4).

2. *Maintain groundwater recharge*. Over 80% of the annual precipitation infiltrates into the soil mantle in most watersheds under natural conditions, with more than half of this rainfall taken up by vegetation and transpired. The rest of this infiltrated water moves deeper into the unsaturated soil and bedrock until it encounters the water table, where it moves down gradient to emerge as springs and seeps, feeding local wetlands and surface streams or enters the aquifers that supply drinking water wells. Without the constant recharge of groundwater, surface stream flows and groundwater in wells would diminish or disappear during drought periods. Certain undeveloped land areas recharge more than others, and protecting these critical recharge areas is important in maintaining the water cycle balance, so understanding how the water balance works on a specific site is critical.

Figure 5-3 Stream bank erosion from development.

Figure 5-4 Increased bankfull frequency with increased imperviousness.

3. *Prevent downstream increases in volume and flooding impacts*. Although site-based rate control measures may help protect the area immediately downstream from a development site, the increased volume of runoff and the prolonged duration of runoff from multiple development sites can actually increase peak flow rates and duration of flood flows from runoff caused by small rain events. Replicating predevelopment runoff volumes for small storms frequently does not address flooding from large storms, but it will reduce substantially the problem of frequent flooding which affects many communities.

4. *Replicate the natural hydrology on site before development*. The objective for stormwater management is to develop a program that replicates the natural hydrologic conditions of watersheds to the extent practicable. However, the very process of clearing the existing vegetation from the site removes the single largest component of the hydrologic regime, evapotranspiration (ET). Unless we replace the ET component, the runoff increase will be substantial and can actually increase the groundwater recharge with infiltration systems. Several of the BMPs described in subsequent sections, such as infiltration, tree planting, vegetated roof systems, and rain gardens, can help replace a portion of the ET function.

Volume Control Guideline

Whatever BMP technologies are found to be suitable for a given development site, the following control guideline is recommended:

> Do not increase the post-development total runoff volume for all storms equal to or less than the 2-year/24-hour event.

The scientific basis for this control guideline is as follows:

- The 2-year event encompasses 95% or more of the annual runoff volume across the state.
- Volume-reduction BMPs based on this standard will provide a storage capacity to help reduce (or eliminate) the increase in peak flow rates for larger runoff events.
- In a natural stream system in mid-Atlantic states, the bankfull stream flow occurs with a period of approximately 1.5 years. If the runoff volume from storms less than the 2-year event are not increased, the fluvial impacts on streams will be significantly lessened.
- The 2-year-frequency rainfall is well defined and weather data are readily accessible for use in stormwater management calculations.

Peak-Rate Control Guideline

Peak-rate control for large storms, up to the 100-year event, is essential to protect against immediate downstream erosion and flooding. Most designs have achieved peak-rate control through the use of detention structures, but peak-rate control can also be integrated into volume-control BMPs in ways that eliminate the need for additional peak-rate control detention systems. Nonstructural BMPs can also contribute to rate control, as discussed later. The control guideline recommended for peak-rate control is

> Do not increase the peak rate of discharge for the 1-year through 100-year events.

Where watershed stormwater planning has been undertaken, hydrologic modeling may have been performed to suggest more stringent release-rate criteria on subbasins within the watershed. As volume-reduction BMPs are incorporated into stormwater management on a watershed basis, release-rate values will warrant reevaluation. Volume-control guidelines will reduce or perhaps even eliminate the increase in peak rate for many storm events.

Water Quality Protection Guideline

The volume control required with the control guideline may result in removal of the major fraction of particulate-associated pollutants from impervious surfaces during most storms. Semipervious surfaces such as *lawnscapes*, subject to continuing fertilization, may generate NPS pollutants throughout a major storm, as may stream banks subjected to severe flows. While infiltration and landscape LID measures are very effective in NPS reduction, if the volume control measures simply overflow during severe storms, they will not achieve the control anticipated. Solutes (such as NO_3) will continue to be transported in runoff throughout the storm, regardless of magnitude, so their control will rely on application reduction and vegetation management programs.

The control guideline will provide water quality control and stream channel protection as well as flood control protection for most storms if the measures drain reasonably well and are sized and distributed adequately. A less stringent volume guideline will not fully mitigate the peak rate for larger storms, and will require the addition of secondary measures for peak-rate control. These secondary measures could provide additional water quality protection. In the event that this secondary measure is added to assure rate mitigation during severe storms, the incorporation of vegetative elements could provide effective water quality controls.

The control guideline recommended for total water quality protection is

> Achieve an 85% reduction in postdevelopment particulate-associated pollutant load (as represented by total suspended solids), an 85% reduction in postdevelopment total phosphorus loads, and a 50% reduction in postdevelopment solute loads (as represented by NO_3-N), all based on postdevelopment land use.

These reductions may be estimated based on the pollutant load for each land use type and the pollutant removal effectiveness of the proposed measures, as shown in other chapters. The inclusion of total phosphorus as a parameter is in recognition of the fact that much of the phosphorus in transit with stormwater is attached to the small (colloidal) soil particles, which are not subject to gravity settlement in conventional detention structures, except over extended periods. With infiltration or vegetative treatment, however, the removal of both suspended solids and total phosphorus should be very high, approaching 100%.

New impervious surfaces such as rooftops, which produce relatively little additional pollutants, can be left out of the water quality impact site evaluation under most circumstances. Rainfall has some latent concentration of nitrate (1 to 2 mg/L) as the result of air pollution, but it would be unreasonable to require removal of this pollutant load from stormwater runoff. The control of nitrate from new development should focus on reduction of fertilizer applications rather than removal from runoff.

When the development plan proposed for a site is measured by type of surface (roof, parking lot, driveway, lawn, etc.), an estimate of potential pollutant load can be made based on the volume of runoff from those surfaces, with a flow-weighted pollutant concentration applied. The total potential non-point source load can then be estimated for the parcel, and the various LID measures, both structural and nonstructural, can be considered for their effectiveness. This method is described in detail in subsequent chapters. In general, the nonstructural measures are most beneficial for the reduction of solutes, with structural LID measures most useful for particulate reduction. Because soluble pollutants, once they are contained in runoff, are extremely difficult to remove, prevention or reduction on the land surface, as achieved through nonstructural measures, is the most effective approach.

Stormwater Standards for Special Areas

The control guideline may have to be modified before applying to special areas around the country. The "special areas" designation includes existing dense urban parcels (served by combined sewers) or highly developed sites with little or no vacant areas, contaminated or brown field sites, and sites situated on old mining lands. These are areas where LID application may be limited.

Legal Implications of Green Infrastructure

Although most research and development of green infrastructure has focused on the technical aspects of various methods and materials, a recent paper by A. P. Dunn raises several issues as to how green infrastructure can alleviate urban poverty and promote healthy communities [13]:

> Green infrastructure is an economically and environmentally viable approach for water management and natural resource protection in urban areas. This article argues that green infrastructure has additional and exceptional benefits for the urban poor which are not frequently highlighted or discussed. When green infrastructure is concentrated in distressed neighborhoods—where it frequently is not—it can improve urban water quality, reduce urban air pollution, improve public health, enhance urban aesthetics and safety, generate green collar jobs, and facilitate urban food security. To make these quality of life and health benefits available to the urban poor, it is essential that urban leaders remove both legal and policy barriers to implementing green infrastructure projects. This Article argues that overcoming

these obstacles requires quantified methods and regulatory reform. Increased public financing and other incentives are also necessary. Furthermore, legal structures that facilitate green solutions must be put in place. Lastly, awareness of green infrastructure solutions among policy makers and the wider public must be enhanced so that our nation's more distressed urban populations may realize the benefits that such solutions yield.

REFERENCES

1. Sullivan, T., 1989. *Environmental Law Handbook*. Government Institutes, The Scarecrow Press, Inc., Rowman & Littlefield Publishing, Landham, MD, 20706. http://www.govinstpress.com.
2. Federal Water Pollution Control Act (FWPCA, PL 92-500), 1972.
3. Marine Protection, Research and Sanctuaries Act (MPRSA, PL 92-536), 1972.
4. National Environmental Policy Act (NEPA PL 91-190), 1969.
5. Pennsylvania Department of Environmental Protection, 2006. *Best Management Practices Manual*. PA DEP, Harrisburg, PA, Dec. 30 (technical consultant, Cahill Associates, Inc.).
6. Kent, T., 1964. *The Urban General Plan*. Chandler Publishing, San Francisco.
7. Parolek, D. G., K. Parolek, and P. C. Crawford, 2008. *Form-Based Codes: A Guide for Planners, Urban Designers, Municipalities, and Developers*. Wiley, Hoboken, NJ.
8. Local Government Commission, 2005. *Ahwahnee Water Principles*. LGC, Sacramento, CA.
9. Corbett, J., et al., July 2006. *The Ahwahnee Water Principles: A Blueprint for Regional Sustainability*. Local Government Commission, Sacramento, CA.
10. California Regional Water Quality Control Board, San Diego Region, 2007. *Waste Discharge Requirements for Discharges of Urban Runoff from the Municipal Separate Storm Sewer Systems (MS4s) Draining the Watersheds of the County of San Diego, the Incorporated Cities of San Diego County, the San Diego Unified Port District, and the San Diego County Regional Airport Authority*. CRWQCB, San Diego, CA.
11. Stein, E. D., and S. Zaleski, Dec. 2005. *Managing Runoff to Protect Natural Streams: The Latest Developments on Investigation and Management of Hydromodification in California*. Technical Report 475. Southern California Coastal Water Research Project, San Diego, CA.
12. Leopold, L. B., M. G. Wolman, and J. P. Miller, 1985. *Fluvial Processes in Geomorphology*. Dover Publications, New York.
13. Dunn, A. P., 2010. *Siting Green Infrastructure: Legal and Policy Solutions to Alleviate Urban Poverty and Promote Healthy Communities*. Environmental Law Programs, Pace University, NY.

Additional Sources

California Regional Water Quality Control Board, Los Angeles Region, Apr. 2008. *Draft Ventura County Municipal Separate Storm Sewer System Permit*. NPDES Permit CAS004002. CRWQCB, Los Angeles, CA.

Governor's Office of Planning and Research, June, 2008. *CEQA and Climate Change: Addressing Climate Change Through California Environmental Quality Act (CEQA) Review*. State of California, Sacramento, CA.

Project Clean Water, July 2008. *Countywide Model SUSMP: Standard Urban Stormwater Mitigation Plan Requirements for Development Applications*. Draft. PCW, San Diego, CA.

van Empel, C., Winter 2008. CEQA and new urbanist development, in *The Environmental Monitor*. Association of Environmental Professionals, Sacramento, CA.

Walsh, S., et al., Dec. 2005. *California Planning Guide: An Introduction to Planning in California*. Governor's Office of Planning and Research, Sacramento, CA.

6

LID DESIGN CALCULATIONS AND METHODOLOGY

6.1 INTRODUCTION TO STORMWATER METHODOLOGIES

The land development process results in significantly greater volumes of runoff and conveys land pollutants to surface waters, but the difficult issue remains as to how to prevent or reduce this impact. Before we consider how best to manage our runoff, we must decide how much of the net increase should or can be reduced or prevented. Since our measurements of the hydrologic cycle are limited to input (rainfall) and output (stream flow), understanding what happens in between in a watershed or catchment is a critical issue.

This is certainly not a new problem, and engineers and hydrologists have developed a number of analytical methods over the past several decades to estimate the amount of runoff produced in a watershed. These various methods are based on the land that comprises the drainage area, and attempts to replicate the complex process of how surface runoff is produced and the multiple pathways followed by each raindrop as it follows the energy gradient downhill. Whatever the algorithm formulated to describe the process, the end result is to estimate the form of the resultant hydrograph that occurs in the receiving stream following rainfall. Since this estimated surface flow hydrograph is expected to mimic the hydrograph observed, all models "calibrate" or adjust input parameters to replicate this energy waveform.

Initially, most of this effort was undertaken to allow the building of structures within the stream or river channel, in the form of culverts, bridges, dams, and

---

Low Impact Development and Sustainable Stormwater Management, First Edition. Thomas H. Cahill.

other hydraulic structures. The key value measured (or estimated) by the hydrologic modeling analysis has been the *peak rate* of flow that will result during a given rainfall at a specific point in the drainage system. The stormwater management strategy of reducing or mitigating runoff volume recommended in this book produces a very different perspective on these various modeling procedures, and begs the question of which method best estimates the change in runoff *volume* resulting from land development.

There have been many methodologies developed to estimate the total runoff volume, the peak rate of runoff, and the stream hydrograph produced from land surfaces under a variety of conditions. In this chapter we describe some of the methods that are most widely used throughout the country. It is certainly not a complete list of procedures, nor is it intended to discourage the use of new and better methods as they become available. There are a wide variety of both public- and private-domain computer models available for performing stormwater calculations, and these models use one or more calculation methodologies to estimate runoff characteristics, following procedures discussed here.

To facilitate a consistent and organized presentation of information throughout the region, assist design engineers in meeting the site design criteria recommended, and help reviewers analyze project data, a series of worksheets are provided in Appendix A for design professionals to complete and submit with their development applications. It should be noted that since the traditional focus of stormwater management in most cities has been related to hydromodification and peak rate, the methods in this chapter for addressing volume, recharge, and quality may be relatively new in some areas of the region.

6.2 EXISTING METHODOLOGIES FOR RUNOFF VOLUME CALCULATIONS

Runoff Curve Number Method

The runoff curve number method, developed by the U.S. Soil Conservation Service [SCS; now the Natural Resources Conservation Service (NRCS)], is perhaps the most commonly used tool in the country for estimating runoff volumes. In this method, runoff is calculated based on precipitation, curve number, watershed storage, and initial abstraction. When rainfall is greater than the initial abstraction (i.e., $P > I_a$), runoff is given by [1]

$$Q_v = \frac{(P - I_a)^2}{(P - I_a) + S}$$

where

Q_v = runoff (in.)
P = rainfall (in.)
I_a = initial abstraction (in.)
S = potential maximum retention after runoff begins (in.)

Initial abstraction (I_a) includes all losses before the start of surface runoff: depression storage, interception, evaporation, and infiltration. I_a can be highly variable, but the SCS has found that it can be approximated empirically by

$$I_a = 0.2S$$

Therefore, the runoff equation becomes (when $P > 0.2S$)

$$Q = \frac{(P - 0.2S)^2}{P - 0.8S}$$

Finally, S is a function of the watershed soil and cover conditions as represented by the runoff curve number (CN):

$$S = \frac{1000}{\text{CN}} - 10$$

Therefore, runoff can be calculated using only the curve number and rainfall. Curve numbers are determined by land cover type, hydrologic condition, antecedent runoff condition (ARC; sometimes referred to as antecedent moisture condition), and hydrologic soil group. Curve numbers for various land covers based on an average ARC for annual floods and $I_a = 0.2S$ can be found in *Urban Hydrology for Small Watersheds* [2], and various other references. Table 6-1 includes some of the more commonly used curve numbers from *Urban Hydrology for Small Watersheds*.

Often, a single, area-weighted curve number is used to represent a watershed consisting of subareas with different curve numbers. Although this approach is

Table 6-1 Commonly Used Curve Numbers

	Curve Numbers for Hydrologic Soil Group			
Cover Type and Hydrologic Condition	A	B	C	D
Runoff Curve Numbers for Urban Areas[a]				
Open spaces (parks, golf courses, cemeteries, etc.)[b]				
Poor condition (grass cover <50%)	68	79	86	89
Fair condition (grass cover 50% to 75%)	49	69	79	84
Good condition (grass cover >75%)	39	61	74	80
Impervious areas				
Paved parking lots, roofs, driveways, etc. (excluding right-of-way) Streets and roads	98	98	98	98
paved; curbs and storm sewers (excluding right-of-way)	98	98	98	98
Paved, open ditches (including right-of-way)	83	89	92	93
Gravel (including right-of-way)	76	85	89	91

(*continued*)

Table 6-1 (*Continued*)

Cover Type	Hydrologic Condition	Curve Numbers for Hydrologic Soil Group A	B	C	D
Runoff Curve Numbers for Other Agricultural Lands[c]					
Pasture, grassland, or range—continuous forage for grazing[d]	Poor	68	79	86	89
	Fair	49	69	79	84
	Good	39	61	74	80
Meadow—continuous grass, protected from grazing and generally mowed for hay	—	30	58	71	78
Brush—brush–weed–grass mixture, with brush the major element[e]	Poor	48	67	77	83
	Fair	35	56	70	77
	Good	30[f]	48	65	73
Woods–grass combination (orchard or tree farm)[g]	Poor	57	73	82	86
	Fair	43	65	76	82
	Good	32	58	72	79
Woods[h]	Poor	45	66	77	83
	Fair	36	60	73	79
	Good	30[f]	55	70	77
Farmsteads—buildings, lanes, driveways, and surrounding lots	—	59	74	82	86

Source: [2, Table 2-2].

[a]Average runoff condition, and $I_a = 0.2S$.

[b]CNs shown are equivalent to those of pasture. Composite CNs may be computed for other combinations of open space cover type.

[c]Average runoff condition, and $I_a = 0.2S$.

[d]Poor: <50% ground cover or heavily grazed with no mulch.
Fair: 50 to 75% ground cover and not heavily grazed.
Good: >75% ground cover and lightly or only occasionally grazed.

[e]Poor: <50% ground cover.
Fair: 50 to 75% ground cover.
Good: >75% ground cover.

[f]Actual curve number is less than 30; use CN = 30 for runoff computations.

[g]CNs shown were computed for areas with 50% woods and 50% grass (pasture) cover. Other combinations of conditions may be computed from the CNs for woods and pasture.

[h]Poor: Forest litter, small trees, and brush are destroyed by heavy grazing or regular burning.
Fair: Woods are grazed but not burned, and some forest litter covers the soil.
Good: Woods are protected from grazing, and litter and brush adequately cover the soil.

acceptable if the curve numbers are similar, if the difference in curve numbers is more than 5, the use of a weighted curve number reduces significantly the estimated amount of runoff from the watershed. This is especially problematic with pervious–impervious combinations: "combination of impervious areas with pervious areas can imply a significant initial loss that may not take place." Therefore, the runoff from different subareas should be calculated separately and then combined or weighted appropriately. At a minimum, runoff volume from pervious and directly connected impervious areas should be estimated separately for storms less than approximately 4 in. [3, 4]. When impervious areas are disconnected effectively from the drainage system, some runoff can be absorbed by pervious surfaces. To account for this, the worksheets in Appendix A include credits for disconnection.

To account for the land development process, all disturbed pervious areas that are not restored using one of the techniques described earilier should be assigned a curve number that reflects a "fair" hydrologic condition as opposed to a "good" condition for post-development volume calculations. For example, lawns should be assigned curve numbers of 49, 69, 79, and 84 for soil groups A, B, C, and D, respectively.

The curve number method is less accurate for storms that generate less than 0.5 in. of runoff, and the SCS recommends using another procedure as a check for these situations. For example, the storm depth that results in 0.5 in. of runoff varies according to the CN; for impervious areas (CN of 98) it is a 0.7-in. storm, for "open space" in good condition on C soils (CN of 74) it is 2.3 in., and for woods in good condition on B soils (CN of 55) it is over 3.9 in. The CN methodology can significantly underestimate the runoff generated from smaller storm events [5]. An alternative method for calculating runoff from small storms is described below.

Recently, some researchers have suggested that the assumption that $I_a = 0.2S$ does not fit the observed rainfall–runoff data nearly as well as $I_a = 0.05S$. The incorporation of this assumption into the curve number method results in a new runoff equation and a new curve number. Woodward [6] describes the new runoff equation and a procedure to convert traditional CNs to new values based on $I_a = 0.05S$. They also describe a plan to implement these changes into all appropriate NRCS documents and computer programs. The most notable differences in runoff modeling with these changes occur at lower curve numbers and lower rainfalls (using the traditional curve number assumption of $I_a = 0.2S$ results in higher initial abstractions and lower runoff volumes under these conditions). When utilized to predict runoff from developed sites in southern California during typical design storms, the difference is likely to be insignificant. Therefore, the standard assumption is used in this chapter. The curve number method, applied with appropriate CNs and the foregoing considerations in mind, is recommended for typical runoff volume calculations and is included in the worksheets.

Small Storm Hydrology Method

The small storm hydrology method (SSHM) was developed to estimate the runoff volume from urban and suburban land uses for relatively small storm events. Other common procedures, such as the runoff curve number method, are less accurate for small storms as described previously. Specifically, the SSHM should be used when the land cover with the lowest curve number for a particular project produces 0.5 inch or less of runoff using the curve number method.

Table 6-2 Small Storm Volumetric Coefficients (R_v) for Urban Runoff

Rainfall (mm)	Rainfall (in.)	Flat Roofs and Large Unpaved Parking Lots	Pitched Roofs and Large Impervious Areas (Large Parking Lots)	Small Impervious Areas and Narrow Streets	Paved Streets	Pervious Areas, Sandy Soils Group A	Pervious Areas, Clayey Soils Groups C and D
1	0.04	0.00	0.25	0.93	0.26	0.00	0.00
3	0.12	0.30	0.75	0.96	0.49	0.00	0.00
5	0.20	0.54	0.85	0.97	0.55	0.00	0.10
10	0.39	0.72	0.93	0.97	0.60	0.01	0.15
15	0.59	0.79	0.95	0.97	0.64	0.02	0.19
20	0.79	0.83	0.96	0.67	—	0.02	0.20
25	1.00	0.84	0.97	0.70	—	0.02	0.21
30	1.25	0.86	0.98	0.74	—	0.03	0.22
38	1.50	0.88	0.99	0.77	—	0.05	0.24
50	2.00	0.90	0.99	0.99	0.84	0.07	0.26
80	3.15	0.94	0.99	0.99	0.90	0.15	0.33
125	4.92	0.96	0.99	0.99	0.93	0.25	0.45

Source: [7].

The SSHM is a straightforward procedure in which runoff is calculated using volumetric runoff coefficients. The runoff coefficients, R_v, are based on extensive field research in the United States and Canada over a wide range of land uses and storm events. The coefficients have also been tested and verified for numerous other U.S. locations. Runoff coefficients for individual land uses generally vary with the rainfall amount—larger storms have higher coefficients. Table 6-2 lists SSHM runoff coefficients for six land use scenarios for storms ranging from 0.04 to 4.92 in. Table 6-3 presents runoff coefficient reduction factors for disconnected impervious surfaces associated with three types of development: strip commercial and shopping center, medium- to high-density residential with paved alleys, and medium- to high-density residential without alleys.

Runoff is simply calculated by multiplying the rainfall amount by the appropriate runoff coefficient (it is important to note that these volumetric runoff coefficients are not equivalent to the peak rate runoff coefficient used in the rational method, discussed below). Since the runoff relationship is linear for a given storm (unlike the curve number method), a single weighted runoff coefficient can be used for an area consisting of multiple land uses. Therefore, runoff is given by

$$Q_v = P \times R_v$$

where

Q_v = runoff (in.)
P = rainfall (in.)
R_v = area-weighted volumetric runoff coefficient

Table 6-3 Reduction Factors to Volumetric Runoff Coefficients (R_v) for Disconnected Impervious Surfaces[a]

Rainfall		Strip Commercial and Shopping Center	Medium- to High-Density Residential with Paved Alleys	Medium- to High-Density Residential Without Alleys
mm	(in.)			
1	0.04	0.00	0.00	0.00
3	0.12	0.00	0.08	0.00
5	0.20	0.47	0.11	0.11
10	0.39	0.90	0.16	0.16
15	0.59	0.99	0.20	0.20
20	0.79	0.99	0.29	0.21
25	1.00	0.99	0.38	0.22
30	1.25	0.99	0.46	0.22
38	1.50	0.99	0.59	0.24
50	2.00	0.99	0.81	0.27
80	3.15	0.99	0.99	0.34
125	4.92	0.99	0.99	0.46

Source: [7].

[a] For low-density residential, use connected values for pervious surfaces with clayey soil from Table 330-13 in [7].

Infiltration Models for Runoff Calculations

Several computer packages offer the choice of using soil infiltration models as the basis of runoff volume and rate calculations. Horton developed perhaps the best-known infiltration equation—an empirical model that predicts an exponential decay in the infiltration capacity of soil toward an equilibrium value as a storm progresses over time [8]. Green and Ampt [9] derived another equation describing infiltration based on physical soil parameters. As the original model applied only to infiltration after surface saturation, Mein and Larson [10] expanded it to predict the infiltration that occurs up until saturation. These infiltration models estimate the amount of precipitation excess occurring over time. The excess must then be transformed to runoff with other procedures to predict runoff volumes and hydrographs.

Urban Runoff Quality Management

Another method for calculating runoff volume is the urban runoff quality management approach [11]. This approach is suitable for planning-level estimates of the size of volume-based BMPs. This approach is based on the translation of rainfall to runoff. The urban runoff quality management approach is based on two regression equations. The first regression equation relates rainfall to runoff. The rainfall to runoff regression equation was developed using two years of data from more than 60 urban watersheds nationwide. The second regression equation relates mean annual runoff-producing rainfall depths to the *maximized water quality capture volume*, which corresponds to the "knee" of the cumulative probability curve. This second regression was based on analysis of long-term rainfall data from seven rain gages representing climatic zones across the country. The maximized water quality capture volume corresponds to approximately the 85th percentile runoff event, and ranges from 82 to 88%.

6.3 EXISTING METHODOLOGIES FOR PEAK-RATE/HYDROGRAPH ESTIMATES

The Rational Method

The rational method has been used for over 100 years to estimate peak runoff rates from relatively small, highly developed drainage areas. Both the standard and modified rational methods may be the most commonly used runoff methods in many states. The peak runoff rate from a given drainage area is given by

$$Q_p = C \times I \times A$$

where

Q_p = peak runoff rate [cubic feet per second (ft^3/sec)]
C = runoff coefficient of the area (assumed to be dimensionless)
I = average rainfall intensity (in./hr) for a storm with a duration equal to the time of concentration of the area
A = size of the drainage area (acres)

The runoff coefficient is usually assumed to be dimensionless because 1 acre-inch per hour is very close to 1 cubic foot per second (1acre-in./hr = 1.008 ft^3/sec). Although it is a simple and straightforward method, estimating both the time of concentration and the runoff coefficient introduces considerable uncertainty in the peak runoff rate calculated. In addition, the method was developed for relatively frequent events, so the peak rate as calculated above should be increased for more extreme events. Because of these and other serious deficiencies, the rational method should be used to predict the peak runoff rate only for very small, highly impervious areas [12].

Although this method has been adapted to include estimations of runoff hydrographs and volumes through the modified rational method, the universal rational hydrograph, the DeKalb rational hydrograph, and so on, these are further compromised by assumptions about the total storm duration and therefore should not be used to calculate volumes related to water quality, infiltration, or capture–reuse.

The NRCS (SCS) Unit Hydrograph Method

In combination with the curve number method for calculating runoff depth, the U.S. Soil Conservation Service (now the NRCS) also developed a system to estimate peak runoff rates and runoff hydrographs using a dimensionless unit hydrograph derived from many natural unit hydrographs from diverse watersheds throughout the country. As discussed below, the NRCS methodologies are available in several public-domain computer models, including the TR-55 computer model [2], the TR-20 computer program [13], and is an option in the U.S. Army Corps of Engineers' hydrologic modeling system [14].

6.4 COMPUTER MODELS

The HEC Hydrologic Modeling System

The U.S. Army Corps of Engineers' Hydrologic Modeling System (HEC-HMS) [14] supersedes HEC-1 as "new-generation" rainfall–runoff simulation software. According to the Corps, HEC-HMS "is a significant advancement over HEC-1 in terms of both computer science and hydrologic engineering." HEC-HMS was designed for use in a "wide range of geographic areas for solving the widest possible range of problems." The model incorporates several options for simulating precipitation excess (runoff curve number, Green and Ampt, etc.), transforming precipitation excess to runoff (SCS unit hydrograph, kinematic wave, etc.), and routing runoff (continuity, lag, Muskingum–Cunge, modified Puls, kinematic wave). HEC-HMS Version 3.2 (April 2008) and supporting materials can be downloaded free at http://www.hec.usace.army.mil/software/hec-hms/download.html.

The SCS/NRCS Models: WinTR-20 and WinTR-55

"The WinTR-20 model is a storm event surface water hydrologic model.... It can be used to analyze current watershed conditions as well as assess the impact of proposed changes (alternates) made within the watershed.... Direct runoff is computed from watershed land areas resulting from synthetic or natural rain events. The runoff is routed through channels and/or impoundments to the watershed outlet" (WinTR-20 User Documentation). TR-20 applies the methodologies found in the hydrology section of the *National Engineering Handbook* [1], specifically the runoff curve number method and the dimensionless unit hydrograph [13]. WinTR-20 Version 1.00 (January 2005) and supporting materials can be downloaded free at http://www.wsi.nrcs.usda.gov/products/W2Q/H&H/Tools_Models/WinTR20.html.

Technical Release 55 (TR-55) was originally published in 1975 as a simple procedure to estimate runoff volume, peak rate, hydrographs, and storage volumes required for peak rate control [15]. TR-55 was released as a computer program in 1986 and work began on a modernized Windows version in 1998. WinTR-55 generates hydrographs from urban and agricultural areas and routes them downstream through channels and/or reservoirs. WinTR-55 uses the TR-20 model for all of its hydrograph procedures [15]. WinTR-55 Version 1.0.08 (January 2005) and supporting materials can be downloaded free at http://www.wsi.nrcs.usda.gov/products/W2Q/H&H/Tools_Models/WinTR55.html.

The Stormwater Management Model

The U.S. Environmental Protection Agency's storm water management model (SWMM) is a dynamic rainfall–runoff simulation model used for single-event or long-term (continuous) simulation of runoff quantity and quality from primarily urban areas. The runoff component of SWMM operates on a collection of subcatchment areas that receive precipitation and generate runoff and pollutant loads. The routing portion of SWMM transports this runoff through a system of pipes, channels, storage and treatment devices, pumps, and regulators. SWMM tracks the quantity and quality of runoff generated within each subcatchment, and the flow rate, flow depth, and quality of water in each pipe and channel during a simulation period comprised of multiple time steps [16, 17].

SWMM is a powerful model capable of simulating areas ranging from a single, uniform subcatchment to the drainage system of an entire city. Although typically not used to evaluate a single development site, the recently released Version 5 is more user-friendly and should promote an increase in use among design professionals.

Rainfall excess is calculated in SWMM by subtracting infiltration (based on Horton [8], Green and Ampt [9], or the NRCS curve number method [3]) and/or evaporation from precipitation. Rainfall excess is converted to runoff by coupling Manning's equation with the continuity equation [18] SWMM Version 5.0.013 (March 2008) and supporting materials can be downloaded at http://www.epa.gov/ednnrmrl/models/swmm/.

The Source Loading and Management Model

The source loading and management model (SLAMM) is designed to provide information about the sources of critical pollutants in urban runoff and the effectiveness of stormwater BMPs for controlling these pollutants. SLAMM was developed primarily as a planning-level model to predict flow and pollutant discharges from a wide variety of development conditions using many combinations of common stormwater BMPs. Development of the model began in the mid-1970s and was supported by the U.S. Environmental Protection Agency, the Wisconsin Department of Natural Resources, and the Ontario (Canada) Ministry of the Environment [19]. Because of their importance for pollutant loading, SLAMM places special emphasis on small storms and utilizes the small storm hydrology method to calculate surface runoff. According to Pitt and Voorhees [19], "SLAMM also calculates correct NRCS curve numbers that reflect specific development and control characteristics. These curve numbers can then be used in conjunction with [other] available urban drainage procedures to reflect the water quantity reduction benefits of stormwater quality controls." The latest version of SLAMM, WinSLAMM version 9.3.1 (July 2008), can be purchased through http://www.winslamm.com/.

Continuous Modeling

The methodology outlined in this chapter is based on single-event calculations using hypothetical design storms (e.g., the 2-year, 24-hour NRCS Type II storm) because they are relatively simple and widely accepted, have been used historically, and are the basis of many of the local standards. However, the advent of better computer models and faster processors has made the continuous simulation of long periods of recorded climate data much more feasible. While continuous simulations generally require extensive precipitation data and much more time to develop, they offer the benefit of analyzing actual long-term conditions rather than one or more hypothetical storms. Legitimate continuous modeling can certainly be used to demonstrate comparable performance to the LID goals recommended in this book. In fact, some jurisdictions in the country are beginning to require continuous simulation to demonstrate compliance with stormwater standards. That being said, the single-event methodology recommended here—with the appropriate assumptions included—is a cost-effective, yet defensible approach for most projects.

6.5 PRECIPITATION DATA FOR STORMWATER CALCULATIONS

In 2004 the National Weather Service's Hydrometeorological Design Studies Center published updated precipitation estimates for much of the United States. NOAA Atlas 14 supersedes previous precipitation estimates such as Technical Memorandum NWS Hydro 35 and Technical Papers 40 and 49 (TP-40 and TP-49) because the updates are based on more recent and expanded data, current statistical techniques, and enhanced spatial interpolation and mapping procedures [20, 21]. NOAA Atlas 14: *Precipitation-Frequency Atlas of the United States* [20], provides estimates of 1-year through 1000-year storm events for durations ranging from 5 minutes to 60 days. These data are available online at http://hdsc.nws.noaa.gov/hdsc/pfds/sa/sca_pfds.html). Users can select precipitation estimates for certain areas in southern California by entering latitude/longitude coordinates or by clicking on an interactive map on the Precipitation Frequency Data Server. These new rainfall estimates are recommended for all applicable stormwater calculations.

The National Weather Service is currently working on completing the precipitation data for the rest of California (http://hdsc.nws.noaa.gov/hdsc/pfds/other/nca_pfds.html). For those areas not yet covered in the recent update, generalized maps for the 6- and 24-hour point precipitation for the return periods of 2, 5, 10, 25, 50, and 100 years can be found in NOAA Atlas 2: *Precipitation Frequency Atlas of the Western United States* [22]. Atlas 2 also provides equations and interpolation diagrams for determining values for durations less than 24 hours and for intermediate return periods and is located at the following link: http://www.nws.noaa.gov/oh/hdsc/PF_documents/Atlas2_Volume11.pdf.

6.6 ACCOUNTING FOR THE BENEFITS OF LID: LINKING VOLUME AND PEAK RATE

The utilization of volume-reduction BMPs and LID practices will obviously reduce or eliminate the amount of storage required for peak-rate mitigation because less runoff is discharged. However, quantifying the peak-rate mitigation benefits of LID can be difficult and cumbersome with common stormwater models/methodologies. In this section we discuss some available tools for quantifying the benefits of LID.

In its *Surface Water and Storm Water Rules Guidance Manual* [23] (available at http://www.mmsd.com/stormwaterweb/index.htm) the Milwaukee Metropolitan Sewerage District (MMSD) describes five methods of accounting for "distributed retention" or LID based on the NRCS unit hydrograph method: truncated hydrograph, scalar multiplication, subtracting retention from rainfall, subtracting retention from runoff, and adjusting CNs (Appendix A of Appendix L: Low Impact Development Documentation). Using the method of subtracting retention from runoff, MMSD developed a spreadsheet model (available from MMSD for $25) called LID Quicksheet 1.2: "Quicksheet allows the user to quickly evaluate various LID features on a development site to reduce ... detention requirements.... LID features included in the Quicksheet include rain gardens, rain barrels, green roofs, cisterns, and permeable pavement." Although Quicksheet seems to be a useful tool, the current version does not appear to account directly for ongoing infiltration during the storm event and therefore may not fully credit LID practices that achieve significant infiltration (the ongoing infiltration volume could be added to the capacity of the LID retention features to make up for this.

6.7 RECOMMENDED LID STORMWATER CALCULATION METHODOLOGY

Stormwater management has traditionally focused on peak-rate control for large storm events (e.g., 10-, 25-, 50-, and 100-year-frequency storms) This low-impact development stormwater book recommends that stormwater management be much more comprehensive, including:

- Total runoff volume
- Water quality
- Peak rate of flow
- Groundwater recharge
- Hydromodification
- Stream channel protection

The site design criteria recommended here are designed to achieve all of these comprehensive stormwater management objectives. It should be noted that control of the peak rate of flow of stormwater runoff remains an important part of stormwater management. This peak-rate criterion is generally based on larger storm events of limited frequency (such as the 100-year storm event). By contrast, the additional elements of stormwater management—volume, groundwater recharge, water quality, and stream channel protection—are based largely on the smaller, more frequent storm events (e.g., 2-year storms and smaller).

Engineers and regulatory officials are generally familiar with the engineering methods and models used to evaluate the rate of runoff for large storm events. There is some consistency in the calculation methodologies used, with the rational and modified rational methods probably being the two most common methodologies applied to estimate rate of runoff. However, local county and city regulations are sometimes based on unreferenced equations. Furthermore, most regulations have been based primarily on peak-rate (release-rate) calculations, which may not be easily integrated with comprehensive stormwater management or LID.

To manage stormwater for volume, groundwater recharge, quality, channel protection, and hydromodification and to integrate both nonstructural and structural BMPs and take reasonable credit for their use, additional or expanded analytical methods are needed. In Appendix A we provide guidance on recommended procedures and methodologies to improve stormwater management, and include worksheets and flowcharts intended to assist in this process.

As stated in the site design process, applicants are urged to meet with the reviewing engineer(s) at the initiation of a project proposal, at which time BMPs to be integrated into the site plan should be discussed. Methodologies to be used in this process should be reviewed and agreed upon. Although not mandated, the following methodological procedure has been developed to provide additional guidance to applicants. These recommended procedures are intended to reduce the extent of judgment that will be required as part of the site design and stormwater management engineering process.

Methods Involving No Routing

Simple Volume Diversion (Off-Line Storage) This is a very simple way to account partially for the effect of volume control BMPs on peak runoff rates. Many computer models have components that allow a "diversion" or "abstraction." The total volume reduction provided by the applicable structural and nonstructural BMPs can be diverted or abstracted from the modeled runoff before it is routed to the detention system (if detention is needed). This approach is very conservative because it does not give any credit to the increased time of travel, fully account for ongoing infiltration, and so on, associated with the BMPs. Even this conservative approach can reduce the detention storage requirements significantly. This method can and should be used in conjunction with the travel time/time of concentration adjustment explained below and does not entail any routing through the volume-reducing BMP(s).

Travel Time/Time of Concentration Adjustment In LID design, time of concentration is the time it takes a drop of water to move from the farthest point in the disturbed area to its discharge from the disturbed area. Time of concentration can be affected by adjusting the length or roughness of natural flow paths and routing through BMPs. If time of concentration is kept constant for pre- and postdevelopment conditions, the peak rate is completely dependent on the volume of surface runoff and can be completely controlled by implementing additional volume control. Repeat steps 5 and 6 for the larger storms and determine if additional volume control can be implemented to control the peak rate.

The use of widely distributed, volume-reducing BMPs can significantly increase the postdevelopment runoff travel time to the point of ultimate site discharge and therefore decrease the peak rate of discharge. The Delaware urban runoff management model calculates the extended travel time through storage elements, even when completely full, to adjust peak flow rates [24]. The extended travel time is essentially the residence time of the storage elements, found by dividing the total storage by the 10-year peak flow rate. This increased travel time can be added to the time of concentration of the area to account for the slowing effect of the volume-reducing BMPs. This can significantly reduce or even eliminate the detention storage required for peak rate control. This method can and should be used with simple volume diversion explained above and does not entail routing through the volume-reducing BMP(s). However, it should not be used with the routing methods described below.

Methods Involving Routing

Composite BMPs with Routing For optimal stormwater management, this manual suggests widely distributed BMPs for volume, rate, and quality control. This approach, however, can be very cumbersome to evaluate in detail with common computer models. To facilitate modeling, similar types of BMPs with similar outlet configurations (weirs, low-flow orifices, etc.) can be combined within the model. For modeling purposes, storage of the combined BMP is simply the sum of the BMP capacities that it represents. A stage–storage–discharge relationship (including ongoing infiltration) can be developed for the combined BMP based on the configuration of the individual systems. The combined BMP(s) can then be routed normally and the results submitted. BMPs that are grouped together in this manner should have similar drainage area/storage volume ratios to ensure that the individual BMPs function properly. This method should not be used in conjunction with the travel time/time of concentration adjustment method described above.

Full BMP Routing, Including Ongoing Infiltration For situations where the methods described above do not apply or for projects with a very few BMPs, these BMPs can be modeled using traditional routing methodologies that take into account ongoing infiltration. Designers are encouraged to select and design BMPs that detain and infiltrate stormwater runoff and release at the predevelopment volume and rate, or as required. See Chapter 7 for design guidance on infiltration and detention BMPs.

6.8 NONSTRUCTURAL BMP CREDITS

The use of nonstructural BMPs is an important part of a project's stormwater management system and deserves to be credited properly in the calculation process. However, nonstructural BMPs must be implemented correctly to be effective. The use of these calculation credits for nonstructural BMPs must be documented fully to the reviewing authorities. A system of checklists and worksheets has been developed to facilitate this process, as summarized in flowcharts A, B, and C in Appendix A.

The following nonstructural BMPs are "self-crediting," in that the use of these BMPs automatically provides a reduction in impervious area and/or stormwater runoff (e.g., curve number) and a corresponding reduction in the required amount of stormwater management. Additionally, use of these BMPs may be affected by other regulations or guidance (master plans, zoning, subdivision, etc.). All of these self-crediting BMPs are strongly encouraged:

- Protect sensitive/special value features.
- Protect/conserve/enhance riparian areas.
- Protect/utilize natural flow pathways.
- Cluster uses.
- Concentrate uses through smart growth.
- Minimize disturbed area.
- Reduce street imperviousness.
- Reduce parking imperviousness.

Although these BMPs are self-crediting and are not elaborated further in these recommended procedures, checklists have been provided and should be completed by applicants when these self-crediting BMPs are being proposed.

The following nonstructural BMPs provide a quantitative stormwater benefit and have been elaborated in Worksheet 3 in Appendix A as part of these recommended procedures:

- Soil compaction minimization in disturbed areas
- Protection of existing trees (part of minimum disturbance)
- Revegetation and reforestation of disturbed areas
- Rooftop disconnection
- Disconnection of impervious areas (nonroof)
- Soil restoration

REFERENCES

1. Natural Resources Conservation Service, 1969–2001. *National Engineering Handbook*, Part 630, Hydrology. Originally published as the *National Engineering Handbook*, Sec. 2. sec4, Hydrology. http://www.wcc.nrcs.usda.gov/hydro/hydro-techref-neh-630.html.

2. Soil Conservation Service, U.S. Department of Agriculture, 1986. *Urban Hydrology for Small Watersheds*, 2nd ed. Technical Release 55. USDA, Washington, DC. http://www.wcc.nrcs.usda.gov/hydro/hydro-tools-models-tr55.html.
3. Natural Resources Conservation Service, 2004. National Water and Climate Center, Hydraulics and Hydrology—Tools and Models Web site, U.S. Department of Agriculture. http://www.wcc.nrcs.usda.gov/hydro/hydro-tools-models.html.
4. Pitt, R., 2003. The Source Loading and Management Model (WinSLAMM): Introduction and Basic Uses. http://unix.eng.ua.edu/~rpitt/SLAMMDETPOND/WinSlamm/Ch1/M1.html#_Introduction#_Introduction.
5. Sorrell, R., July 2003. *Computing Flood Discharges for Small Ungaged Watersheds*. Geological and Land Management Division, Michigan Department of Environmental Quality, Lansing, MI.
6. Woodward, D. E., R. H. Hawkins, R. Jiang, A. T. Hjelmfelt, J. A. Van Mullenm, and D. Q. Quan, June 2003. Runoff curve number method: examination of the initial abstraction ratio. Presented at the World Water and Environmental Resources Congress, Philadelphia.
7. Pitt, R. E., Apr. 1997. Section 5, Small Storm Hydrology, in *Stormwater Quality Management Through the Use of Detention Basins: A Short Course on Stormwater Detention Basin Design Basics by Integrating Water Quality with Drainage Objectives*. University of Minnesota Continuing Education and Extension, Minneapolis, MN.
8. Horton, R. E., 1940. An approach toward a physical interpretation of infiltration capacity. *Proceedings of the Soil Science Society of America*, vol. 4, pp. 399–417.
9. Green, W. H., and G. A. Ampt, 1911. Studies on soil physics: 1. The flow of air and water through soils. *Journal of Agricultural Sciences*, vol. 4, p. 1124.
10. Mein, R. G., and C. L. Larson, 1973. Modeling infiltration during a steady rain. *Water Resources Research*, vol. 9, no. 2, pp. 334–394.
11. Clar, M., et al., Sept. 2004. *Stormwater Best Management Practice Design Guide*, Vol. 2. EPA/600/R-04/121A. U.S. Environmental Protection Agency, Washington, DC. http://www.epa.gov/nrmrl/pubs/600r04121/600r04121.htm.
12. Huber, W., et al., July 2006. *BMP Modeling Concepts and Simulation*. EPA/600/R-06/033. U.S. Environmental Protection Agency, Washington, DC. http://www.epa.gov/nrml/pubs/600r06033/toc.pdf.
13. Soil Conservation Service, U.S. Department of Agriculture, 1992. *TR-20 Computer Program for Project Formulation Hydrology*. USDA, Washington, DC.
14. U.S. Army Corps of Engineers, 2001. *Hydrologic Modeling System (HEC-HMS) User's Manual*, Version 2.1. USACE, Davis, CA.
15. Natural Resources Conservation Service, 2002. NRCS, Department of Agriculture, Washington, DC.
16. U.S. Environmental Protection Agency, 2007. Storm Water Management Model (SWMM) Version 5.0.011 Web site. http://www.epa.gov/ednnrmrl/swmm/#A.
17. Rossman, L., 2004. *Storm Water Management Model User's Manual, Version 5.0*. National Risk Management Research Laboratory, U.S. Environmental Protection Agency, Cincinnati, OH. http://www.epa.gov/ednnrmrl/swmm/#A.
18. James, W., W. Huber, R. Dickinson, R. Pitt, W. R. James, L. Roesner, and J. Aldrich, 2003. *User's Guide to SWMM*. Computational Hydraulics International, Guelph, Ontario, Canada.

19. Pitt, R., and J. Voorhees, 2000. The Source Loading and Management Model (SLAMM): A Water Quality Management Planning Model for Urban Stormwater Runoff. http://unix.eng.ua.edu/~rpitt/SLAMMDETPOND/WinSlamm/Main WINSLAMM_book.html.
20. Bonnin, G., D. Todd, B. Lin, T. Parzybok, M. Yekta, and D. Riley, 2003. *NOAA Atlas 14: Precipitation-Frequency Atlas of the United States*, Vol. 1. National Weather Service, Silver Spring, MD.
21. National Weather Service, Hydrometeorological Design Studies Center, 2004. Current Precipitation Frequency Information and Publications Web site, National Oceanic and Atmospheric Administration. http://www.nws.noaa.gov/ohd/hdsc/currentpf.htm.
22. National Oceanic and Atmospheric Administration, 1973. *NOAA Atlas 2: Precipitation Atlas of the Western United States*, Vol. 11. National Weather Service, Silver Spring, MD.
23. Milwaukee Metropolitan Sewerage District, 2001. *Surface Water and Storm Water Rules Guidance Manual*. MMSD, Milwaukee, WI.
24. Lucas, J., 2004. *The Delaware Urban Runoff Model*. Delaware Department of Natural Resources and Environmental Conservation, Dover, DE. http://www.swc.dnrec.delaware.gov/SedimentStormwater.htm.

Additional Sources

Claytor, R. A., and T. R. Schuler, 1996. *Design of Stormwater Filtering Systems*. Center for Watershed Protection, Silver Spring, MD.

Huff, F. A., and J. R. Angel, 1992. *Rainfall Frequency Atlas of the Midwest*. Bulletin 71. Midwestern Climate Center and Illinois State Water Survey. MCC Research Report 92-03. http://www.sws.uiuc.edu/pubdoc/B/ISWSB-71.pdf.

Linsley, R., J. Franzini, D. Freyberg, and G. Tchobanoglous, 1992. *Water-Resources Engineering*, 4th ed. Irwin McGraw-Hill, New York.

Mecklenburg County, 2007. *Mecklenburg County BMP Design Manual*, Chap. 4. http://www.charmeck.org/Departments/StormWater/Contractors/BMP+Standards+Manual.htm.

New Jersey Department of Environmental Protection, 2004. *New Jersey Stormwater Best Management Practices Manual*. NJ DEP, Trenton, NJ.

Reese, S., and J. Lee, 1998. *Summary of Groundwater Quality Monitoring Data (1985–1997) from Pennsylvania's Ambient and Fixed Station Network (FSN) Monitoring Program: Selected Groundwater Basins in Southwestern, Southcentral and Southeastern Pennsylvania*. Bureau of Water Supply Management, Pennsylvania Department of Environmental Protection, Harrisburg, PA.

7

DESIGN OF LID SYSTEMS

7.1 NONSTRUCTURAL MEASURES

In Chapter 4, the site design process for low-impact development (LID) begins with measures that will reduce or avoid the impacts of conventional site design, as illustrated in Figure 4-7, step 6. The basic concept of nonstructural LID measures is to prevent the problem created by land development, including runoff from new impervious surfaces, loss of vegetation, and compaction of the soil mantle, before they happen. These nonstructural measures are applied at the site planning stage and include three basic concepts: (1) reduction of the impervious surface required to achieve a desired building program, (2) limitation of the site disturbance necessary to meet the design program, and (3) planning the site to achieve the program within less space. These measures have taken a number of different forms, based on the imagination of the design team and the willingness of both the owner and the local government to allow flexibility in the site planning process. Examples of successful applications of each approach are provided here not as a standard but as successful examples of the principle.

Impervious Surface Reduction

When we build a building or set of buildings for a given function, we have traditionally surrounded it with an impervious surface on which to store or move our transportation vehicles, augmented by similar surfaces to connect the vehicles

Low Impact Development and Sustainable Stormwater Management, First Edition. Thomas H. Cahill.

with the structure for pedestrian movements. The first question to be answered by the design team is: Can we design the minimum surface to allow this structure to function as required? The second question is: Can we provide the same function in a box with a smaller footprint in multiple levels? The third question goes to the building materials used: Can we provide vehicle storage and pedestrian movement on surfaces that are not impervious? Finally, the issue of vehicle storage begs the question, Do we need to access the site by this mode, or can high-volume transport provide all or a portion of the demand?

Vehicle storage alternatives compare vertical systems, such as parking garages, with at-grade storage in open parking lots. The determining factor here is cost; parking garage costs run about $20,000 to $23,000 per space in current designs, while at-grade sites are an order of magnitude less, in the range of $2,000 to $3,000 per space. Of course, many, if not most, at-grade lots in urban centers are "placeholders" for future buildings, when the value of land will warrant redevelopment.

Limitation of Site Disturbance

On undeveloped sites, the approach is to disturb only as much of a parcel as required to complete the design program. The limitation of disturbance suggests that every undeveloped site has some portion where the impact of development will result in the loss of important resources, such as woodlands, stream valleys, or sensitive habitats. Very little open land remains in most communities that does not have such constraints, especially in the more active development markets. It seems that all the open fields comprised of well-drained soils and within striking distance of existing infrastructure (roadways, water supply, wastewater treatment) have long since been developed. Therefore, the initial step in limiting site disturbance is to identify those portions of the parcel that can be considered environmentally sensitive, and limit site development to the balance of the land.

The second concept is one of site planning, where we try to group the planned units, be they residential or commercial, in tighter, more efficient clusters. The efficiency results in the reduction of infrastructure required to serve structures in this type of configuration. This type of site design is often combined with impervious surface reduction and has been advocated with a program called *sustainable site design*, which has produced some very attractive residential developments.

Site Design with Less Space

A number of residential site designs over the past two decades have proposed model concepts that are basically smaller units as single or multiple structures (twins, townhouses, quads, etc.) situated in groupings that encourage pedestrian movement, and include centralized recreational spaces (pools, parks, play fields, or most important, golf courses). Well-known examples can be encountered in every part of the country.

This concept has more to do with how you build on a parcel of land rather than where you build. For each building envelope, we must confine soil excavation and compaction to the area surrounding (and supporting) the structure. The limitation of grading and earthwork begins by considering carefully how one will enter and exit, how the structure is related to other contiguous buildings in terms of form and function, and how the building will be situated to achieve optimal scenic views of the surrounding community and environment. Historic site design has included extreme examples of structures that totally disregarded this concept in order to create a structure that became the center of community interest, power, or religion. From temples to castles to cathedrals, all supporting structures in the community were designed to draw attention and focus on the central structure.

However, this type of site design has long since vanished into history (although we still honor many remnant examples) and been replaced with community designs that are generally bland and boring. Current site designs have generally followed a practice of total site disturbance and alteration of the land surface (and subsurface) in order to achieve some preconceived notion of aesthetics.

7.2 STRUCTURAL MEASURES

A number of structural measures have been developed for stormwater management and are presently in practice (Table 7-1). Volume reduction is the primary design objective in LID designs, with pollutant transport reduction of equal importance, and rate mitigation an element of all of the design concepts presented. The three basic volume-reduction methods of infiltration, evapotranspiration and capture–reuse are reflected in the four system designs presented in this chapter, with a detailed description of each design approach. In an effort to avoid replication of data, the basic designs are very similar, and all of the measures listed are variations on the basic theme. Since evapotranspiration is the essential element of all vegetation-based designs, two variations are illustrated here.

The general heading "bioremediation" has been in use for some time and has come to describe any stormwater management system that includes vegetation as a primary element. The term evolved from the use of vegetation in retention basins to accomplish some nutrient uptake and allow ET to reduce the standing pool in the basin. Since initial detention basin designs were intended to be mowed lawn surfaces with no plantings allowed, this concept was a radical departure during the late 1980s, but offered great benefit in the correct setting.

The two bioremediation designs described in this book use vegetation in two very different applications: one within the site at grade (rain garden) and the second applying vegetation on the structure proper (vegetated roof). By their nature they include both volume reduction and pollutant reduction, but with very different design parameters.

Table 7-1 LID Structural Measures

Design Concept	Structural Measure	Method
Volume reduction and pollutant removal	Porous pavement with infiltration bed	Infiltration
	Infiltration basin	Infiltration
	Infiltration bed (grass)	Infiltration
	Infiltration swale, trench	Infiltration
	French drain, dry well	Infiltration
	Rain garden	Bioremediation, Infiltration
	Bioremediation bed	Bioremediation
	Vegetated filter strip	Bioremediation
	Vegetated swale	Bioremediation
	Planter box	Bioremediation, evapotranspiration
	Constructed wetland	Bioremediation, evapotranspiration
	Vegetated roof	Evapotranspiration
	Rooftop capture, storage	Capture/potable use
	Rooftop capture, storage	Capture/nonpotable use
	Site capture, storage irrigation use	
Pollutant removal only	Manufactured systems	Filtration
	Sand filter	Filtration
	Wet pond	Bioremediation
Rate mitigation only	Detention basin (dry)	Detention
	Vegetated basin	Detention, limited bioremediation
Other	Vegetation, soil restoration	

7.3 PERVIOUS PAVEMENT WITH AN INFILTRATION OR STORAGE BED

Pervious pavement consists of a permeable surface course underlain by a storage reservoir consisting of a uniformly graded aggregate bed (Figure 7-1) or premanufactured structural stormwater units. The bed is typically laid on uncompacted soil to facilitate rainfall infiltration. The surface course may consist of pervious bituminous asphalt, pervious concrete, various types of pervious pavers, or other types of pervious structural materials [1, 2].

Types of Porous Pavement

Porous pavement wearing surfaces have been constructed of various materials, intended for different types of service, from vehicle traffic (Figure 7-2) to pedestrian use (Figure 7-3):

- Pervious bituminous pavement
- Pervious concrete

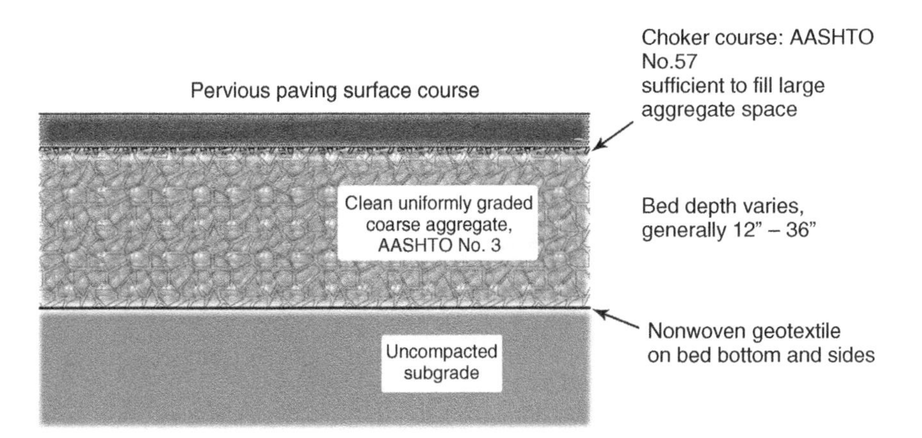

Figure 7-1 Porous pavement section.

Figure 7-2 Porous pavement in traffic application, Portland, Oregon.

- Pervious paver blocks
- Reinforced turf and gravel
- "Clear" binder pavements

The depth of pavement surface varies, but the subsurface design should be the same, with the storage and infiltration bed depth a function of the soil infiltration rate and the inflow surface volume. At present, the longest record of construction experience is available with the porous AC pavement, in use since the mid-1970s [3, 4].

Figure 7-3 Porous concrete sidewalk, State College, Pennsylvania.

Description and Function

A pervious pavement system consists of a pervious surface course underlain by a storage reservoir placed on uncompacted subgrade to facilitate stormwater infiltration (Figure 7-4). The storage reservoir may consist of a stone bed of uniformly graded and clean-washed course aggregate, 1.5 to 2.5 in. in size, with a void space of at least 40% or other premanufactured structural storage units. The pervious pavement may consist of pervious bituminous asphalt, pervious concrete, pervious pavers, or other types of pervious structural materials. Stormwater drains through the surface course, is held temporarily in the voids of the stone bed (Figure 7-5), and then slowly ex-filtrates into the underlying, uncompacted soil mantle.

The stone bed can be designed with an overflow control structure so that during large storm events peak rates are controlled, and at no time does the water level rise to the pavement level (Figure 7-6). A layer of nonwoven geotextile filter fabric separates the aggregate from the underlying soil, preventing the migration of fines into the bed. The bed bottoms should be level and uncompacted to allow for even and distributed stormwater infiltration. If new fill is required, it should consist of additional stone and not compacted soil. It is recommended that a fail-safe factor be built into the system in the event that the pervious surface should be affected adversely and suffer reduced performance. Many designs incorporate a riverstone/rock edge treatment or inlets tied directly to the bed so that the stormwater system will continue to function despite the performance of the pervious pavement surface.

Pervious pavement is well suited for parking lots (Figures 7-7), walking paths (Figure 7-8), sidewalks, playgrounds (Figure 7-9), plazas, tennis courts, and similar uses. Pervious pavement can be used in driveways if the homeowner is aware of the stormwater functions of the pavement and willing to maintain it. Pervious pavement roadways have seen wider application in Europe and Japan

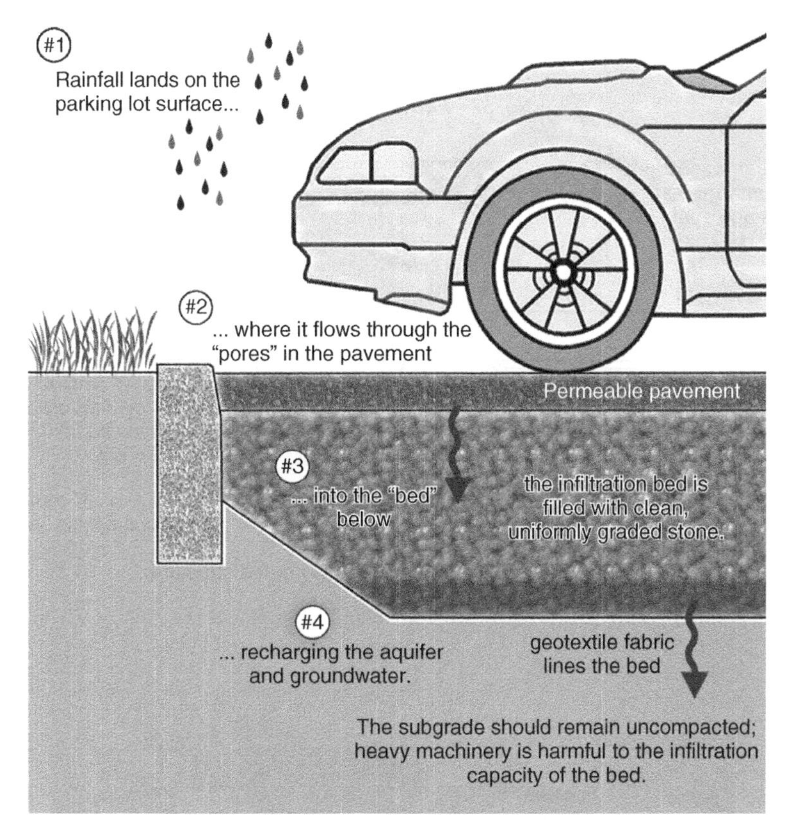

Figure 7-4 Porous pavement design.

Figure 7-5 Stone storage and infiltration bed.

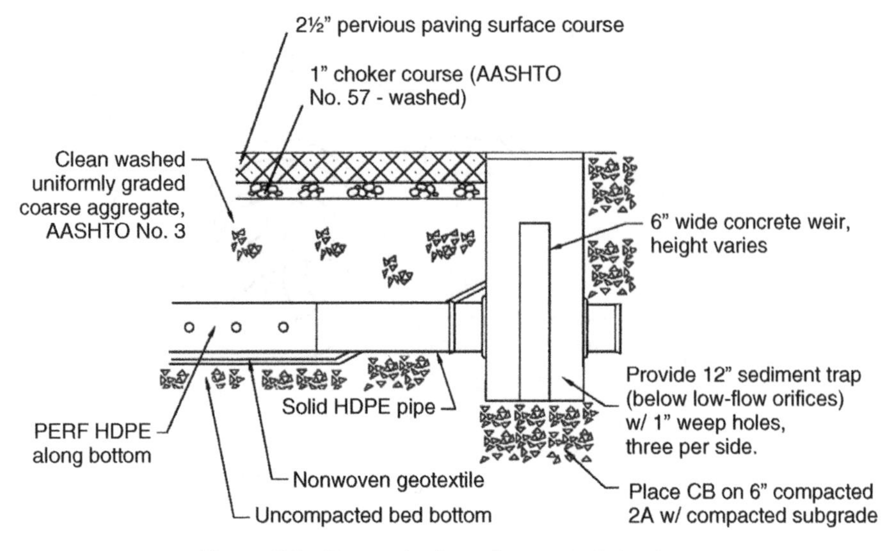

Figure 7-6 Storage bed overflow control structure.

Figure 7-7 Porous pavement next to standard pavement in parking lot.

than in the United States, although at least one U.S. system has been constructed successfully (in Arizona). In Japan and the United States, the application of an open-graded asphalt pavement of 1 in. or less on roadways has been used to provide lateral surface drainage and prevent hydroplaning, but these are applied over impervious pavement on compacted subgrade. This application is not considered a stormwater BMP.

Figure 7-8 Porous AC pavement as a walking path.

Figure 7-9 Porous pavement in playground.

Properly installed and maintained pervious pavement has a significant lifespan, and existing systems that are more than 20 years old continue to function. Because water drains through the surface course and into the subsurface bed, freeze–thaw cycles do not tend to adversely affect pervious pavement. Pervious pavement is most susceptible to failure difficulties during construction, and therefore it is important that the construction be undertaken in such as way as to prevent:

- Compaction of underlying soil
- Contamination of stone subbase with sediment and fines
- Tracking of sediment onto pavement
- Drainage of sediment-laden waters onto pervious surface or into constructed bed

Staging, construction practices, and erosion and sediment control must all be taken into consideration when using pervious pavements.

When designed properly, pervious pavement systems provide effective management of stormwater volume and peak rates. The storage reservoir below the pavement surface can be sized to manage both direct runoff and runoff generated by adjacent areas, such as rooftops. Because the stone bed provides storage, outlet structures can be designed to manage peak rates with the use of weir and orifice controls. A well-designed system can infiltrate the majority of frequent small storms on an annual basis while providing peak-rate control for design storms up to and including the 100-year-frequency storm event.

Studies have shown that pervious systems have been very effective in reducing contaminants, such as total suspended solids, metals, and oil and grease (Table 7-2). Because pervious pavement systems often have zero net discharge of stormwater for frequent small storms, they provide effective water quality

Table 7-2 Water Quality Benefits of Infiltration

System Component	Mechanism(s)	Contaminants Retained/Reduced
Porous pavement	Filtration and adsorption	Total suspended solids (TSS), heavy metals, hydrocarbons, COD, and deicing salt (less required, more retained) (*Note: Maintenance by vacuuming is required!*)
Infiltration bed	Filtration, adsorption, settling, microbial biodegradation	TSS, metals, and hydrocarbons, plus total organic carbon, COD, nitrogen
Shallow soil	Filtration, adsorption, ion exchange, microbial biodegradation, conversion, and uptake (only with high plant activity)	Metals and hydrocarbons, including PAHs
Deeper soil	Filtration, adsorption, ion exchange, conversion, and uptake (only with high plant activity)	Metals and hydrocarbons, plus organics and bacteria; very low risk of groundwater contamination

control. Although the pervious surface and underlying soils below the storage bed allow filtration of most pollutants, care must be taken to prevent infiltration in areas where toxic or contaminated materials are present in the underlying soils or within the stormwater itself. When designed, constructed, and maintained according to the guidelines, pervious pavement with underlying infiltration systems can dramatically reduce both the rate and volume of runoff, recharge the groundwater, and improve water quality.

Pervious Bituminous Asphalt

Pervious bituminous asphalt pavement was first studied in the early 1970s at the Franklin Institute in Philadelphia [5] and consists of standard bituminous asphalt in which the fines have been screened and reduced, allowing water to pass through small voids. Pervious asphalt is placed directly on the stone subbase in a single 3.5- to 4-in. lift that is lightly rolled to a finish depth of 2.5 to 3 in.

Because pervious asphalt is standard asphalt with reduced fines, it is similar in appearance to standard asphalt [6]. Recent research in open-graded mixes for highway applications has led to additional improvements in pervious asphalt through the use of additives and higher-grade binders. Pervious asphalt is suitable for use in any climate where standard asphalt is appropriate [7].

Pervious Portland Cement Concrete

Pervious portland cement concrete, or pervious concrete, was developed by the Florida Concrete Association in the 1990s [8, 9] and has seen the most widespread application in Florida and other southern areas. Like pervious asphalt, pervious concrete is produced by substantially reducing the number of fines in the mix in order to establish voids for drainage. Like other types of pervious pavements, pervious concrete should always be underlain by a stone subbase designed for stormwater management and should never be placed directly onto a soil subbase.

Whereas pervious asphalt is very similar in appearance to standard asphalt, pervious concrete has a coarser appearance than its conventional counterpart (Figure 7-10) and a trowel-swept finish cannot be achieved. Care must be taken during placement to avoid overworking the surface and creating an impervious layer. Pervious concrete has proven to be an effective stormwater management BMP. Another potential advantage of pervious concrete is the option of introducing color to the mix. The industry now offers a variety of hues and tints that can allow a pervious concrete installation to better integrate with its adjacent landscape. Additional information pertaining to pervious concrete, including specifications, is available from the Florida Concrete Association and the National Ready Mixed Concrete Association [10].

Pervious Paver Blocks

Pervious paver blocks [11] consist of interlocking units (often concrete) that provide some portion of surface area that may be filled with a pervious material such

Figure 7-10 Porous concrete produces a course surface.

Figure 7-11 Pervious paver blocks (infiltration through the joint spaces).

as gravel (Figure 7-11). These units are often very attractive and are especially well suited for plazas, patios, small parking areas, and the like. There are also products available that provide a fully permeable surface through the use of plastic rings or grids filled with gravel. A number of manufactured products are available, including (but not limited to) Turfstone, UNI Eco-stone, EP Henry ECO I Paver, Checkerblock, GravelPave, Netlon Gravel Pavement Systems, and Permapave.

As products are always being developed, the designer is encouraged to evaluate the benefits of various products with respect to the specific application. Many paver manufacturers recommend compaction of the soil and do not include a drainage or storage area, and therefore they do not provide optimal stormwater management benefits. A system with a compacted subgrade will not provide significant infiltration.

Reinforced Turf

Reinforced turf consists of interlocking structural units that contain voids or areas for turf grass growth and are suitable for traffic loads and parking (Figure 7-12). Reinforced turf units may consist of concrete or plastic and are underlain by a stone and/or sand drainage system for stormwater management. Reinforced turf is excellent for applications such as fire-access roads, overflow parking, and occasional-use parking (such as at religious facilities and athletic facilities). Reinforced turf is also an excellent application to reduce the required standard pavement width of paths and driveways that must occasionally provide for emergency vehicle access.

Although both plastic and concrete units perform well for stormwater management and traffic needs, plastic units tend to provide better turf establishment and longevity, largely because the plastic will not absorb water and diminish soil moisture conditions. A number of products are available and the designer is encouraged to evaluate and select a product suitable to the design in question. A number of manufactured products are available, including (but not limited to) Grasspave, Geoblock, Grassy Pave, Geoweb, and Netlon Turf Pavement Systems.

Figure 7-12 Reinforced turf.

Other Porous Surfaces

Other proprietary products are now available which are similar to pervious asphalt and concrete but which utilize clear binders so that the beauty of the natural stone is visible, allowing an extremely aesthetically pleasing look. Some of these products are not suitable for vehicular traffic, but the material strength varies by product. The use of clear binder allows the designer the versatility of utilizing different colored aggregates to suit the application and appearance desired. Typical applications include tree pits, walkways, plazas, and playgrounds. A number of products are available on the market, including (but not limited to) Addapave TP and Flexipave.

Potential Applications

Pervious pavements can be used in residential, commercial, institutional, and industrial applications in both urban and suburban environments. The use of pervious pavements has also been widely applied in retrofit situations when existing standard pavements are being replaced. Care must be taken when using pervious pavements in industrial and commercial applications where pavement areas are used for material storage or the potential for surface clogging is high, due to pavement usage.

Pervious Pavement Walkways (Concrete and Asphalt)

Pervious pavement, both asphalt and concrete, has also been used in walkways and sidewalks. These installations typically consist of a shallow (8 in. minimum) aggregate trench that is sloped to follow the surface slope of the path. In the case of steeper surface slopes, the aggregate infiltration trench may be "terraced" into level reaches in order to maximize its infiltration capacity, at the expense of additional aggregate.

Rooftop and Impervious Area Connections

Pervious pavement systems are often used to provide total site stormwater management, where rooftops and other impervious surfaces are drained into the infiltration bed below the pavement surface (Figure 7-13). This can be an effective means to manage stormwater for a development site while reducing land disturbance for stormwater BMPs. Proper sediment pretreatment for runoff from adjacent areas must be considered to prevent clogging of the storage bed. Typical pretreatment can be achieved by the use of properly maintained inlet sediment traps and/or water quality inserts (or other filter devices). It is recommended that direct surface sheet flow conveyance of large impervious areas to the pervious pavement surface be avoided. High-sheet-flow loading to pervious pavement surfaces can lead to premature clogging of the pavement surface. To avoid this, it is recommended that adjacent impervious areas be drained and conveyed to the infiltration bed via inlets and/or trench drains with proper sediment pretreatment.

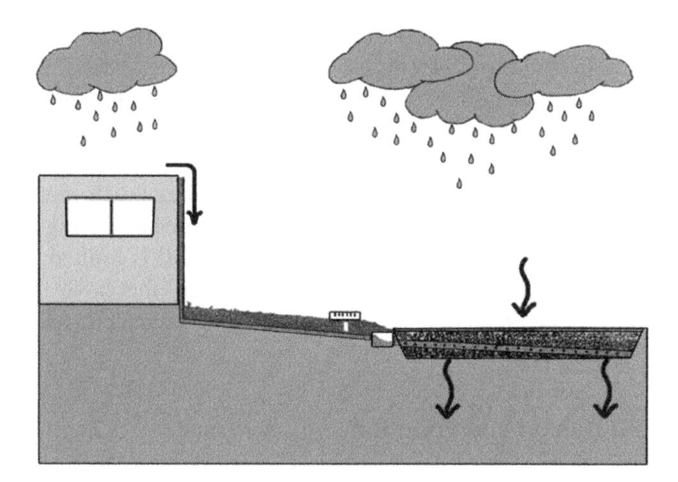

Figure 7-13 Rooftop connection to storage and infiltration bed.

Water Quality Mitigation

Pervious pavement systems are effective in reducing such pollutants as total suspended solids (TSS), metals, and oil and grease (see Table 7-2). Both the pervious pavement surface and the underlying soils below the infiltration bed allow pollutant filtration. When pervious pavement systems are designed to capture and infiltrate runoff volumes from small storm events, or what is often referred to as the *water quality volume*, they provide very high pollutant reductions because there is little if any discharge of runoff carrying the highest pollutant loads. Because pervious pavement systems require pretreatment of TSS when adjacent areas are allowed to drain to them, reduction of TSS and other particulates is typically high. Pervious pavement systems will also provide limited treatment of dissolved pollutants such as nitrates. Typical ranges of pollutant reduction efficiencies for pervious pavements based on available literature and sampling data are as follows: TSS, 65 to 100%; TP, 80 to 90%; TN, 30 to 65%; and NO_3, 30%. Pretreatment is recommended for TSS if contiguous areas drain to pervious pavement. For information on calculating pollutant removal benefits provided by pervious pavements, see Chapter 6.

7.4 BIOREMEDIATION

The use of vegetation in stormwater management systems evolved from vegetated detention basins (originally described as bioretention [12], where a portion of the runoff was retained to support vegetative growth, usually wetland species, in a relatively impervious basin. While detention and retention structures are no longer considered an adequate stormwater management method, the use of vegetation has proved very effective, and partially replicates the natural system removed or destroyed in the land development process. This LID method utilizes vegetation as part of the runoff volume-reduction process, based on the natural transpiration of plant materials as they return rainfall to the atmosphere. The method also provides an infiltration bed as part of the planted area, so they offer two ways to reduce runoff. When the soil mantle is highly impervious, the role of vegetation and evapotranspiration (ET) becomes much more important. Plants provide large surface areas where evaporation occurs during much of the year in most locations, but in arid regions where rainfall is limited, the ET may actually represent a resource depletion. The LID designs that have evolved using vegetation as essential elements include rain gardens, vegetated roofs, and urban designs that are labeled "green streets."

Rain Garden: Design and Function

Rain gardens [13, 14] manage stormwater by pooling water within a planting area and then allowing the water to infiltrate the garden, where underlying soils permit infiltration. In addition to managing runoff volume and mitigating peak discharge rates, this process filters suspended solids as well as pollutants from stormwater runoff. They can be implemented in small, residential applications or as part of a management strategy in larger applications (Figure 7-14).

A rain garden can be designed as a typical landscape feature, which also improves water quality while reducing runoff quantity. Rain gardens can be integrated into a site with a high degree of flexibility and can complement other structural management systems, including porous pavement parking lots, infiltration trenches, and nonstructural stormwater management measures [15].

Figure 7-14 Commercial site rain garden.

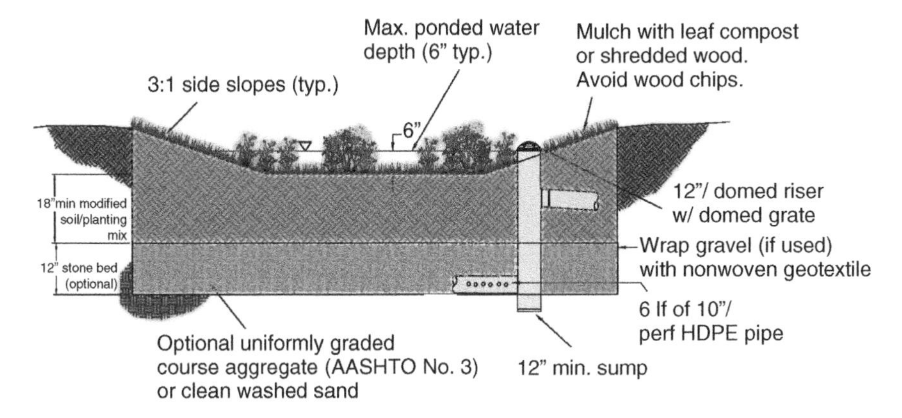

Figure 7-15 Rain garden schematic.

Rain garden vegetation serves to filter (improve water quality) and transpire (reduce runoff quantity) stormwater. The plants absorb some pollutants, while microbes associated with the plant roots and soil can also break down pollutants. In addition to filtering pollutants, the soil medium allows storage and (where feasible) infiltration of stormwater runoff, providing volume control. In addition, engineered soil media may serve as a bonding surface for nutrients to enhance removal of this pollutant source in runoff. Additional treatment capacity is provided by a surface mulch layer, which traps sediments that can have great pollutant loads. A botanically diverse rain garden, with a variety of plant species and types, can help build a system that tolerates insect pests, diseases, pollution, and climatic stresses [16].

Figure 7-15 illustrates a schematic diagram of a rain garden that is a more technically engineered structure, designed to complete specific stormwater management goals. Pond depth, soil mixture, infiltration bed, perforated underdrains, domed risers, and positive overflow structures may be designed in urban environments according to the specific stormwater management functions that are required.

Primary Components of a Rain Garden System

Pretreatment (may be necessary to help prevent clogging)

- Sediment removal through a vegetated buffer strip, cleanout, stabilized inlet, water quality inlet, or sediment trap prior to runoff entry into the rain garden

Flow entrance

- Varies with site (e.g., parking island versus residential applications)
- Water may enter via an inlet (e.g., flared end section) or trench drain
- Sheet flow into the facility over grassed areas or level spreader

- Curb cuts with grading for sheet flow entrance
- Roof leaders with direct surface connection
- Entering velocities must be nonerosive where concentrated runoff enters the rain garden—use erosion control mats or blankets and/or inlet energy dissipaters such as rocks or splash blocks

Ponding area

- Provides temporary surface storage of runoff
- Provides evaporation for a portion of runoff
- Allows sediment to settle
- Depth no more than 6 to 18 in. for aesthetics, functionality, and safety

Plant material

- Absorbs stormwater through transpiration
- Root development creates pathways for infiltration
- Bacteria component of the plant–soil community helps create healthy soil structure with water quality benefits
- Can improve site appearance
- Provides habitat for animals
- Should be native plant species
- Ensure that plants can tolerate snowmelt chemicals, if applicable (at high elevations)
- Should be placed according to water and saturation tolerance

Surface mulch or organic layer

- Acts as a filter for pollutants in runoff
- Protects underlying soil from drying and eroding
- Reduces the likelihood of weed establishment
- Provides a medium for biological growth, decomposition of organic material, and adsorption and bonding of heavy metals
- Wood mulch should be shredded for easier decomposition—compost or leaf mulch is preferred

Planting soil/volume storage bed

- Makes water and nutrients available to plants
- Enhances biological activity and encourages root growth
- Provides storage of stormwater by the voids within the soil particles
- Provides surface for adsorption of nutrients in runoff

Figure 7-16 Rain garden overflow structure.

Positive overflow

- Provides for the direct discharge of excess runoff during large storm events when the subsurface and surface storage capacity is exceeded
- Examples of outlet controls include domed risers, inlet structures, weirs, and similar devices are shown in Chapter 8

Domed riser. A domed riser may be installed in a rain garden to ensure positive, controlled overflow from the system (see Figure 7-16). Once water ponds to a specified depth, it will begin to flow into the riser through a grate, which is typically domed to prevent the riser from being clogged by debris.

Inlet structure. An inlet structure may also be installed in a rain garden to ensure positive, controlled overflow from the system (Figure 7-16). Once water ponds to a specified depth, it will begin to flow into the inlet.

Generally, a rain garden system is a depression in the ground, with plants and a surface mulch layer, which provides for the storage and infiltration of relatively small volumes of stormwater runoff, often managing stormwater on a lot-by-lot basis. This use of many small stormwater controls versus one large detention area promotes the low-impact development goal of decentralized stormwater treatment. If greater volumes of runoff must be managed or stored, a rain garden system can be designed with an expanded subsurface infiltration bed, or each rain garden can be increased in size. Typically, the ratio of impervious area draining to the rain garden to the rain garden area should not exceed 5 : 1, and the total impervious area draining to a single system should not be more than 1 acre [17, 18].

The most common variation includes a gravel or sand bed underneath the planting bed. The original intent of this design, however, is to perform as a filter BMP utilizing an underdrain with subsequent discharge. Should a designer decide to use a gravel or sand bed for volume storage under the planting bed, additional design elements such as separation fabric and piping should be considered. Plant species composition generally depends on how often water is expected to pond in the rain garden. For southern California, species will probably need to be drought-tolerant plants that can handle occasional inundation during the rainy season.

Various methods are available to allow runoff to enter rain gardens. Pretreatment of runoff should be provided where sediment or pollutants entering the rain garden may cause concern or decreased functionality of the BMP, such as from parking areas; rooftop runoff may need little or no pretreatment. Soil erosion control mats or blankets, plus energy dissipaters, should be used where concentrated runoff flows from impervious areas into the rain garden.

Curbs can be used to direct runoff from an impervious surface along a gutter to a low point where it flows into the rain garden through a curb cut. Curb cuts may be depressed curbs, as shown in Figure 7-17, or may be full-height curbs with openings cast or cut into them. Trench drains can accept runoff from impervious surfaces and convey it to a rain garden (Figure 7-18). The trench drain may discharge to the surface of the rain garden or may connect directly to an aggregate infiltration bed beneath. Figure 7-19 depicts a linear rain garden feature along a highway. Runoff is conveyed along the concrete curb (bottom of photo) until it reaches the end of the gutter, where it spills into the vegetated area.

Figure 7-17 Rain garden curb cut.

Figure 7-18 Trench drain inlet to rain garden.

Figure 7-19 Highway right-of-way rain garden.

In parking lots for commercial, industrial, institutional, and other uses, stormwater management and green space areas are limited. In these situations, rain gardens for stormwater management and landscaping may provide multiple benefits.

7.5 VEGETATED ROOF SYSTEMS

An extensive vegetated roof cover is a veneer of vegetation that is grown on and covers an otherwise conventional flat or pitched roof, endowing the roof ($<30^\circ$ slope) with hydrologic characteristics that more closely match surface vegetation than the roof. The overall thickness of an extensive vegetated roof may range from 2 to 6 in. and may contain multiple layers, including waterproofing, synthetic insulation, non-soil-engineered growth media, fabrics, and other synthetic components. Vegetated roof covers can be optimized to achieve water quantity and water quality benefits. Through the appropriate selection of materials, even thin vegetated covers can provide significant rainfall retention and detention functions. This system is built in a single-medium design, a dual-media design, or a dual-media design with a synthetic layer.

This technology has been used successfully in Europe, especially Germany, since the 1970s [19] and has become the standard method of reducing the volume of runoff generated in urban areas. Most cities in Germany have enacted ordinances that require that any new or reconstructed building include a vegetated roof system to reduce the runoff to the city sewer system, many of which are combined. Those cities that were effectively destroyed and rebuilt following World War II, such as Stuttgart, have almost every rooftop covered with vegetation (Figure 7-20). Other countries have utilized this system on major commercial buildings, such as the Schiphol International Airport in Amsterdam (Figure 7-21).

Figure 7-20 Center City, Stuttgart, Germany. Vegetated rooftops are required.

Figure 7-21 Reception building, Schiphol Airport, Amsterdam, Netherlands.

Figure 7-22 Chicago City Hall vegetated roof.

The technology was initially brought to the United States largely through the pioneering work of one engineer, Charles Miller [20], in the late 1990s. His projects are numerous and nationwide, but one of the best known is the green roof on the Chicago City Hall (Figure 7-22), which has proven to be a highly visible example of the system. Many of the concepts and illustrations in the following pages are drawn from his experience. Of course, like every new idea,

Figure 7-23 Oldest vegetated roof (1650), northern Sweden, seventeenth century.

we often find that someone in the distant past had the same idea, although for a somewhat different purpose (Figure 7-23).

Design and Function

Extensive vegetated roof covers (Figure 7-24) are 6 in. or less in depth and are typically intended to achieve a specific environmental benefit, such as rainfall runoff mitigation, or to help reduce building energy needs for cooling. Although some installations are open to public access, most extensive vegetated roof covers offer at most visual access for the public. To make vegetated roofs practical for installation on conventional roof structures, lightweight materials are used in the preparation of most engineered media. Developments in the last 40 years that have made these systems usable include (1) recognition of the value of vegetated covers in restoring nearly open-space hydrologic performance on impervious surfaces, (2) advances in waterproofing materials and methods, and (3) recognition of the multiple environmental benefits provided by vegetated roofs.

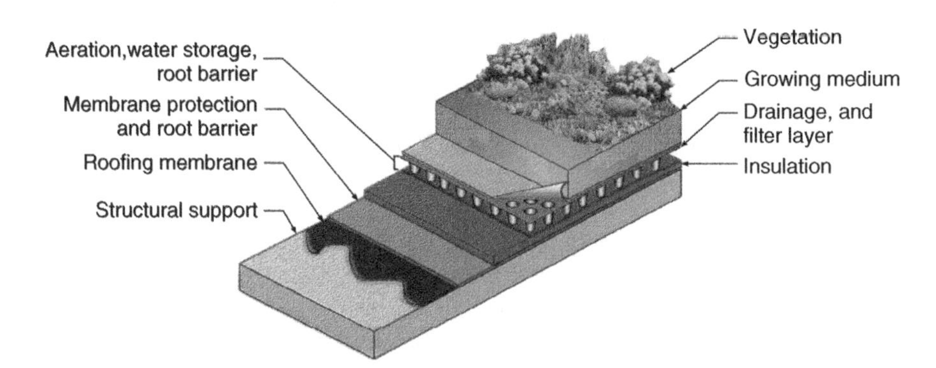

Figure 7-24 Vegetated roof system: extensive.

Vegetated roof covers that are 10 in. deep, or deeper, are referred to as *intensive* vegetated roof covers. These are more familiar in the United States and include many above-structure urban landscaped plazas. Intensive assemblies can also provide substantial environmental benefits, but are intended primarily to achieve aesthetic and architectural objectives. These types of systems are considered roof gardens and are not to be confused with the simple *extensive* design. This LID summary focuses on extensive vegetated roofs, which are more cost-effective as a stormwater management measure.

Design Elements of a Vegetated Roof System

- Engineered media should have a high mineral content; media for extensive vegetated roof covers are typically 85 to 97% nonorganic.
- Two to six inches of non-soil-engineered medium; assemblies that are 4 in. and deeper may include more than one type of engineered media.
- Vegetated roof covers intended to achieve water quality benefits generally should not be fertilized.
- Irrigation may be required for plant survival and optimal functioning of the vegetated roof for stormwater management in southern California.
- Internal building drainage, including provisions to cover and protect deck drains or scuppers, must anticipate the need to manage large rainfall events without inundating the cover.
- Assemblies planned for roofs with pitches steeper than 2 : 12 (9.5°) must incorporate supplemental measures to ensure stability against sliding.
- The roof structure must be evaluated for compatibility with the maximum predicted dead and live loads. Typical dead loads for wet extensive vegetated covers range from 8 to 36 pounds per square foot (lb/ft^2). Live load is a function of rainfall retention. For example, 2 in. of rain equals 10.4 lb/ft^2 of live load.
- The waterproofing must be resistant to biological and root attack. In many instances a supplemental root-fast layer is installed to protect the primary waterproofing membrane from plant roots.
- Standards and guidelines (in English) for the design of green roofs are available from FLL, a European nonprofit trade organization. In the United States, new standards and guidelines are in development by the American Society for Testing and Materials.

Types of Vegetated Roof Systems

Most extensive vegetated roof covers fall into one of three categories:

1. Single medium with synthetic underdrain layer
2. Dual media
3. Dual media with synthetic retention/detention layer

All vegetated roof covers require a premium waterproofing system. Depending on the waterproofing materials selected, a supplemental root-fast layer may be required to protect the primary waterproofing membrane from plant roots. Insulation, if included in the roof covering system, may be installed either above or below the primary waterproofing membrane. Most vegetated roof cover systems can be adapted to either roofing configuration. In the descriptions that follow, the assemblies refer to the conventional configuration, in which the insulation layer is below the primary waterproofing.

All three extensive roof cover variations can be installed with or without irrigation. Irrigated assemblies are strongly recommended for use in arid environments, given the region's climate and precipitation patterns. Irrigation will ensure plant survival in the thin, porous growth medium, especially during dry periods. In addition, dry-season irrigation allows the plants to provide the evaporative cooling benefits to a building when it is most needed. Some assemblies are installed in traylike modules to facilitate installation, especially in confined locations. These assemblies may be difficult to use with subsurface irrigation systems.

Single-medium assemblies (Figure 7-25) are commonly used for pitched-roof applications and for thin and lightweight installations. These systems typically incorporate very drought-tolerant plants and utilize coarse engineered media with high permeability. A typical profile would include the following layers:

- A waterproofing membrane
- A root barrier (optional, depending on the root-fastness of the waterproofing)
- A semirigid plastic geocomposite drain or mat (typical mats are made from nonbiodegradable fabric or plastic foam)

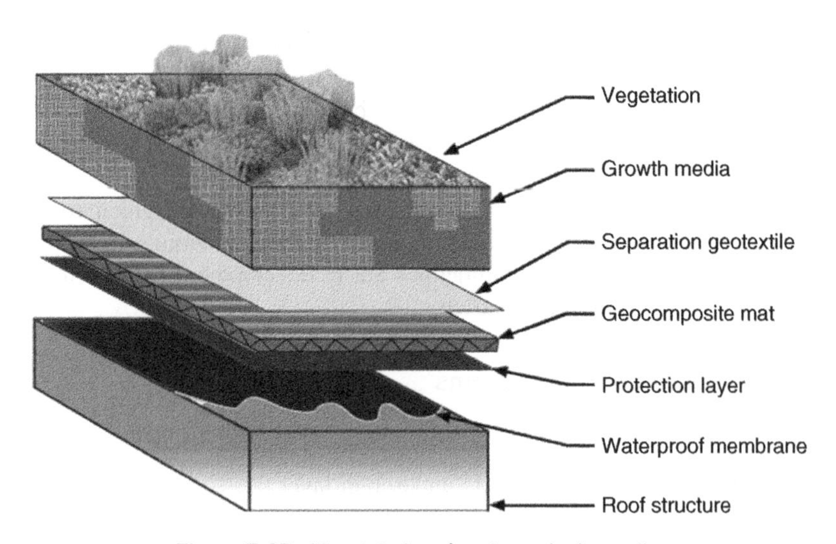

Figure 7-25 Vegetated roof system: single media.

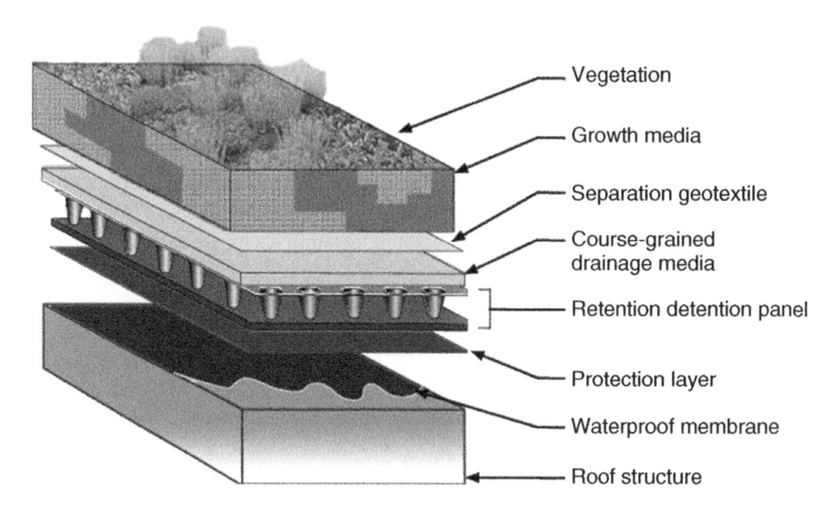

Figure 7-26 Vegetated roof system: dual media.

- A separation geotextile
- An engineered growth medium
- A vegetation layer

Pitched-roof applications may require the addition of slope bars, rigid slope stabilization panels, cribbing, reinforcing mesh, or a similar method of preventing instability and sliding. Flat roof applications with mats as foundations typically require a network of perforated internal drainage conduit to enhance drainage of percolated rainfall to the deck drains or scuppers.

Dual-media assemblies (Figure 7-26) utilize two types of nonsoil media. In this case a finer-grained medium with some organic content is placed over a base layer of coarse lightweight mineral aggregate. They do not include a geocomposite drain. The objective is to improve the drought resistance by replicating a natural alpine growing environment in which sandy topsoil overlies gravelly subsoil. These assemblies are typically 4 to 6 in. thick and include the following layers:

- A waterproofing membrane
- A protection layer
- A coarse-grained drainage medium
- A root-permeable nonwoven separation geotextile
- A fine-grained engineered growth medium layer
- A vegetation layer

These assemblies are suitable for roofs with pitches less than or equal to about 1.5 : 12 (7.1°). Large vegetated covers will generally incorporate a network of perforated internal drainage conduit. Dual-media systems are ideal for base irrigation methods.

Dual Media with a Synthetic Retention Layer

Dual media assemblies with a synthetic retention layer introduce plastic panels with cuplike receptacles on their upper surface (i.e., a modified geocomposite drain sheet). The panels are in-filled with coarse lightweight mineral aggregate. The cups trap and retain water. They also introduce an air layer at the bottom of the assembly. A typical profile would include:

- A waterproofing membrane
- A felt fabric
- A retention/detention panel
- A coarse-grained drainage medium
- A separation geotextile
- A fine-grained growth medium layer
- A vegetation layer

These assemblies are suitable on roof with pitches less than or equal to 1 : 12 (4.8°). Due to their complexity, these systems are usually 5 in. or deeper. If required, irrigation can be provided via surface spray or midlevel drip.

Potential Applications

Assuming that a roof is structurally adequate to support a vegetated roof load (Figure 7-27), and that the roof slope is less than 20% (although not absolutely flat, as this may hinder proper drainage), vegetated roofs can potentially be used for most buildings. Vegetated roofs can be employed for all land uses; the greatest stormwater benefits will be realized in highly developed sites where the building footprint comprises a significant portion of the site, and for sites with poorly drained soils. Buildings on sites with greater perviousness may also use vegetated roofs, although it may be less expensive to use at-grade LID measures for stormwater management purposes; vegetated roofs can be used in conjunction with at-grade measures. Vegetated roofs may be used to help achieve benefits other than mitigating stormwater runoff. These include evaporative cooling (which lowers building energy use for HVAC systems and helps minimize the urban heat island effect), noise reduction inside the building, increased roof longevity, habitat provision, and improved aesthetic value.

7.6 CAPTURE–REUSE

There are many designs and manufactured systems whose function is to intercept and store runoff, primarily from rooftops [21–24]. The stormwater that is contained can be used for irrigation or other nonpotable water needs without disinfection. For potable systems, the stored rainfall is further filtered and disinfected (by an ultraviolet tube) with in-line units prior to consumption. This design

(A)

(B)

Figure 7-27 Vegetated roof: retrofit installation.

varies from simple rain barrels at residential installations to larger storage tanks and various treatment and delivery hardware for large commercial installations.

Capture–reuse is a term that encompasses the practice of collecting rainwater in a storage system and using it for potable or nonpotable purposes. Other terms for this BMP include storage–reuse, rainwater harvesting, and rainwater catchment system. This type of structural BMP can reduce potable water needs while simultaneously reducing stormwater discharge volumes. This technique captures rooftop runoff via small or large containers (Figure 7-28). Storage units

Figure 7-28 Rooftop runoff capture system.

are available in a wide range of capacities, for above- or below-grade installation, depending on local climate and potential for freezing conditions.

Internal storage is also possible if the roof drain system uses scuppers and internal storage space is available. Large commercial or industrial structures are most likely to include this type of roof drainage and accommodate storage units within the structure. If only external storage is available, heating elements can be used to prevent cold weather problems. For storage structures with public water augmentation, a backflow prevention valve is required on the public service. All reuse systems must be pressurized for internal distribution, 30- to 50-psi gage. A major consideration when designing capture–reuse systems is the quality of the water to be used.

The recent demonstration site at the San Diego County Operations Center in Kearney Mesa included the opportunity to collect and analyze four samples of runoff from the roof of the large warehouse building adjacent to the porous pavement installation (Figure 7-29). The chemistry of that rainfall runoff is shown in Table 7-3 and includes comparison with two relevant drinking water standards; the San Diego County Source Water Protection Guidelines (SWPG, 2004), which apply to the protection of water quality in the regional reservoir system as "raw" water sources, and the federal EPA drinking water standards for potable supply. Although hardly a comprehensive data set for regional decision making with respect to the application of this LID technology, it does provide an interesting "snapshot" of the potential chemistry of rainfall discharging from a relatively old roof material in the region. Other rooftop chemical sampling [25, 26] from other parts of the United States offers additional insight into the issue and suggests that local conditions, such as fossil fuel sources in the atmosphere and the proximity of major highways, can introduce a number of trace metals and other chemicals into the rainfall that might prove of concern if the flow is not filtered and disinfected.

Figure 7-29 Rooftop runoff sampling system, San Diego warehouse.

- Storm events equal to or less than the 2-year-frequency rainfall are captured with multiple storage structures, depending on the configuration of the roof, type of vertical drainage (internal or external), and number of collection points.
- Total rooftop runoff capture, with overflow for large storm events, ideally to surface infiltration beds.
- Storage capacity must be sufficiently large to capture runoff from subsequent rainfall events.
- Consider site topography, placing structure upgradient to eliminate pumping needs. For internal storage systems, site topography is not critical.
- If sufficient storage capacity is available, connect overflow to landscaping and the irrigation system.
- Continual use will result in minimum storage capacity.

Rain Barrels and Cisterns

Commonly, rooftop downspouts are connected to a rain barrel (container) that collects runoff and stores water until needed for a specific use (Figure 7-30). Rain barrels are often used at individual homes where water is reused for garden irrigation, including landscaped beds, trees, or other vegetated surfaces. Other uses include commercial and institutional [27, 28]. Most residential models have a minimum capacity of 132 gallons.

A cistern is a container or structure that has a much greater storage capacity than that of a rain barrel [29, 30]. Cisterns may be comprised of prefabricated

Table 7-3 Summary of Results of Chemical Analysis of Stormwater Runoff from Warehouse Roof, San Diego County Left

	Roof			
SanConstituent	12/7/07	12/19/07	1/5/08	1/23/08
Hardness (mg/L)	5	9.4	2.4	15
SSC (mg/L)	9.78	43.6	6.63	25.5
>63 μm	4.09	13	3.03	10.2
<63 μm	5.69	30.6	3.6	15.2
TSS (mg/L)	7.3	18	18	22
COD (mg/L)	4.1	14	7	17
DOC (mg/L)	2.1	3.8	1.5	1.1
Total P	0.029	0.089	0.032	0.078
Ortho-P (mg/L)	0.013	0.015	0.0074[a]	0.018
Ammonia-N (mg/L)	0.23	0.067[a]	0.071[a]	0.16
TKN (mg/L)	0.48	0.95	0.32	0.84
Nitrate-N (mg/L)	0.29	0.37	0.049[a]	0.082J
Total cadmium (μg/L)	0.49	0.33	0.079[a]	0.26
Diss. cadmium (μg/L)	0.38	0.35	0.076[a]	0.12[a]
Total copper (μg/L)	3.7	4.5	1.2	4.9
Diss. copper (μg/L)	2.1	4.4	0.85	0.78
Total lead (μg/L)	3.4	1.7	0.66	24
Diss. lead (μg/L)	0.46	1.3	0.24	0.37
Total zinc (μg/L)	83	81	53	79
Diss. zinc (μg/L)	72	79	42	34
Calculated values[b]				
$Oil\&grease_{COD}$(mg/L)	3.9	4.2	4	4.3
$Oil\&grease_{DOC}$(mg/L)	0.74	1.2	0.57	0.46

Source: Data from Kinnetic Labs, Santa Cruz, CA, 2008.
[a]The value is an estimate.
[b]An equipment malfunction prevented the autosampler from sampling.

materials (fiberglass, plastics, steel, or concrete) or constructed in place (concrete, brick, or other materials) and can be installed underground or on the surface. The storage size of cisterns can range from 200 to 10,000 gallons (Figure 7-31). Greater storage can be provided in below-grade structures, but multiple tanks are usually more efficient.

Rain barrels are best used in areas that are adjacent to landscaped beds or gardens, since this BMP is used primarily to supplement irrigation. Rain barrels can be used on residential properties, but can also be used at schools and campuses, commercial offices, and any other area where highly aesthetic landscaping features are important. Cisterns can be used in any land use where a significant water need exists. Cisterns are used to either supplement gray water needs (toilet flushing or some other sanitary sewer use), for irrigation, or for potable use following disinfection and fine filtration.

Figure 7-30 Rain barrel system.

Figure 7-31 Large cistern system.

Figure 7-32 Vertical storage unit, Stuttgart office building.

Vertical Storage

A vertical storage element is a structure designed to contain a large volume of stormwater drained from a large impervious area. These structures are stored on the surface and can be integrated into commercial sites where water needs may be higher. This type of structure is often found throughout cities in Europe (Figure 7-32). Stormwater can also be stored under hardscaped elements (such as paths and walkways) through the use of structural plastic storage units, such as RainTank or alternative manufactured storage products, and can supplement on-site irrigation needs.

A vertical storage element is the largest of the capture–reuse containers, and therefore its use is a function of drainage area and water needs. Vertical structures are best used for intensive irrigation needs or even fire suppression requirements. Designing a capture–reuse system in which the cistern is under a hardscaped structure is best used in institutional or commercial settings. This type of subsurface storage is larger and more elaborate and typically requires pumps to connect to the irrigation system.

References and Additional Sources

Porous Asphalt Concrete

1. Cahill, T., 1993. *Porous Pavement with Underground Recharge Beds: Engineering Design Manual*. Cahill Associates, West Chester, PA.
2. Ferguson, B., 2005. *Porous Pavements*. CRC Press, Boca Raton, FL.
3. Cahill, T., 1994. A second look at porous pavement/underground recharge. *Watershed Protection Techniques*, vol. 1, pp. 76–78.
4. Adams, M., 2003. Porous asphalt pavement with recharge beds: 20 years and still working. *Stormwater*, vol. 4, pp. 24–32.
5. Thelen, E., and L. F. Howe, 1978. *Porous Pavement*. Franklin Institute Press, Philadelphia.
6. Cahill, T., M. Adams, and C. Marm, 2003. Porous asphalt: the right choice for porous pavements. *Hot Mix Asphalt Technology*, Sept.–Oct.
7. Backstrom, M., 1999. Porous Pavement in a Cold Climate. Licentiate thesis. Lulea University of Technology, Lulea, Sweden. http://epubl.luth.se.

Porous Cement Concrete

8. Florida Concrete and Products Association, no date. *Construction of a Portland Cement Pervious Pavement*. FCPA, Orlando, FL.
9. Paine, J. E., 1990. *Stormwater Design Guide: Portland Cement Pervious Pavement*. Florida Concrete and Products Association, Orlando, FL.
10. Tennis, P. D., M. L. Leming, and D. J. Akers, 2004. *Pervious Concrete Pavements*. Portland Cement Association, Skokie, IL, and National Ready Mixed Concrete Association, Silver Spring, MD.

Porous Pavers

11. Smith, D. R., 2001. *Permeable Interlocking Concrete Pavements: Selection, Design, Construction, Maintenance*, 2nd ed. Interlocking Concrete Pavement Institute, Washington, DC.

Rain Garden References

12. *Prince George's County Bioretention Manual*, 2002. Department of Environmental Resources, Prince George's County, MD.
13. Clar, M. et al., Oct. 2007. Rethinking bioretention design concepts. Presented at the Pennsylvania Stormwater Management Symposium, Villanova, PA.

14. *Rain Garden Design*. University of Wisconsin–Extension and Wisconsin Department of Natural Resources, 2002.
15. *Pennsylvania Stormwater Best Management Practices Manual*, Dec. 2006. Pennsylvania Department of Environmental Protection, Harrisburg, PA.
16. Rain Gardens of West Michigan. http://www.raingardens.org.
17. *Minnesota Stormwater Manual*, 2006. Minnesota Pollution Control Agency, St. Paul, MN.
18. Southeastern Oakland County Water Authority. http://www.socwa.org/lawn_and_garden.htm.

Vegetated Roofs

19. Cantor, S., 2008. *Green Roofs in Sustainable Landscape Design*. W. W. Norton & Co., NYC, New York. ISBN-10 0393731685.
20. Miller, C., 2007. Roofmeadows, Inc.,7135 Germantown Ave., Philadelphia, PA 19119. www.roofmeadow.com.

Rain Harvesting

21. Phillips, A. A., Ed., Mar. 2003. *Water Harvesting Guidance Manual*. Prepared for the City of Tucson, Department of Transportation, Stormwater Section, Tucson, AZ.
22. McDonough, W., and M. Braungart, 2002. *Cradle to Cradle: Remaking the Way We Make Things*. North Point Press, San Francisco, CA.
23. Harvested rainwater guidelines, Secs. 1.0, 2.0, and 3.0 in Sustainable Building Sourcebook. http://www.greenbuilder.com.
24. City of Austin, TX. Rainwater Harvesting. http://www.ci.austin.tx.us/greenbuilder/fs_rainharvest.htm.
25. Mason, Y., A. A. Ammann, A. Ulrich, and L. Sigg, 1999. Behavior of heavy metals, nutrient, and major components during roof runoff infiltration. *Environmental Science and Technology*, vol. 33, pp. 1588–1597.
26. Van Metre, P. C., and B. J. Mahler, 2003. *The contribution of particles washed from rooftops to contaminant loading to urban streams*. *Chemosphere*, vol. 52, pp. 1727–1784.
27. Sands, K., and, T. Chapman, 2005. *Rain Barrels: Truth or Consequences*. Milwaukee Metropolitan Sewerage District, Milwaukee, WI.
28. *Rain Barrel Program*. City of Vancouver, Engineering Services, Water and Sewers, Vancouver, BC, Canada.
29. Cisterns/rainwater harvesting systems, in Technologies and Practices: Plumbing and Water Heating. http://www.advancedbuildings.org.
30. Black Vertical Storage Tanks by Norwesco. http://www.precisionpump.net/storagetanksystems.htm.

Additional Sources

Porous Asphalt Pavement

Hossain, M., L. A. Scofield, and W. R. Meier, Jr., 1992. Porous pavement for control of highway runoff in Arizona: performance to date. *Transportation Research Record*, vol. 1354, pp. 45–54.

Jackson, N., 2003. *Porous Asphalt Pavements*. Information Series 131. National Asphalt Pavement Association, Lanham, MD.

Kandhal, P. S., 2002. *Design, Construction, and Maintenance of Open-Graded Asphalt Friction Courses*. Information Series 115. National Asphalt Pavement Association, Lanham, MD.

Kandhal, P. S., and R. B. Mallick, 1998. *Open-Graded Asphalt Friction Course: State of the Practice*. Report 98-7. Auburn University National Center for Asphalt Technology, Auburn, AL.

Kandhal, P. S., and R. B. Mallick, 1999. *Design of New-Generation Open-Graded Friction Courses*. Report 99-2. Auburn University National Center for Asphalt Technology, Auburn, AL.

Mallick, R. B., P. S. Kandhal, L. A. Cooley, Jr., and D. E. Watson, 2000. *Design, Construction and Performance of New-Generation Open-Graded Friction Courses*. Report 2000-01. Auburn University National Center for Asphalt Technology, Auburn, AL.

Tappeiner, W. J., 1993. *Open-Graded Asphalt Friction Course*. Information Series 115. National Asphalt Pavement Association, Lanham, MD.

Rain Gardens

Lawrence Technological University Research. Rain Gardens: A Household Way to Improve Water Quality in Your Community. http://www.ltu.edu/stormwater/bioretention.asp.

Wild Ones Natural Landscapers. http://www.for-wild.org/.

Capture/Reuse

City of San Diego, CA, 2004. *Source Water Protection Guidelines*.

CSIRO, Land and Water. Urban Water Reuse: Frequently Asked Questions (South Australia)

Portland, OR Code Guide Office of Planning and Development Review, Mar. 2001. *Rainwater Harvesting*, ICC RES/34/1 and UPC/6/2.

U.S. EPA National Pollutant Discharge Elimination System. *Post-Construction Storm Water Management in New Development and Redevelopment: On-Lot Treatment*.

Xiao, Q., E. G. McPherson, and J. R. Simpson, 2007. *Hydrologic Processes at the Residential Scale*. Hydrologic Sciences Program, UC Davis, Center for Urban Forest Research, USDA Forest Service, Davis, CA.

8

STRUCTURAL MEASURES: CONSTRUCTION, OPERATION, AND MAINTENANCE

8.1 POROUS PAVEMENT SYSTEMS

Construction

Although there is no "standard" design for porous pavement systems, experience over the past 30 years has evolved a strategy for design that was outlined in Chapter 7. When a design progresses to a final set of construction drawings, supporting construction guidance and specifications are also required. Rather than providing a "model," the framework for these specifications is provided in this chapter.

1. The overall site should be evaluated for potential pervious pavement or infiltration areas early in the design process, as effective pervious pavement design requires consideration of grading.
2. The soil through which infiltration is to occur must have physical and chemical characteristics (such as appropriate cation-exchange capacity, organic content, clay content, and infiltration rate) that are adequate for proper infiltration duration and treatment of urban runoff for the protection of groundwater beneficial uses.
3. Orientation of the parking bays along the existing contours will significantly reduce the need for cut and fill.

Low Impact Development and Sustainable Stormwater Management, First Edition. Thomas H. Cahill.

4. Pervious pavement and infiltration beds should not be placed on areas of recent fill or compacted fill. Any grade adjustments requiring fill are made using stone subbase material. Areas of historic fill (>5 years) may be considered for pervious pavement, but permeability testing is essential.

5. For the bed to infiltrate, the bed bottom must not be compacted. The stone subbase should be placed in lifts and lightly rolled according to the specifications.

6. During construction, the excavated bed may serve as a temporary sediment basin or trap. This will reduce overall site disturbance. If used as a temporary sediment basin or trap, the bed is excavated to an elevation of at least 12 in. higher than the designed bed bottom elevation. Following construction and site stabilization, sediment is removed and the final grades established.

7. Bed bottoms must be level or nearly level. Sloping bed bottoms will lead to areas of ponding and reduced stormwater distribution within the bed.

8. All systems are designed with an overflow system. Water within the subsurface stone bed should never rise to the level of the pavement surface. Inlet boxes can be used for cost-effective overflow structures. All beds should empty within 72 hours.

9. While infiltration beds are typically sized to handle the increased volume from the more frequent small storms, they must also be able to convey and mitigate the peak rates of the less frequent, more intense storms (such as the 100-year storm). Water-level control in the beds is usually provided in the form of an outlet control structure. A modified inlet box with an internal weir and low-flow orifice is a common type of control structure (see Figure 7-6). The specific design of these structures may vary, depending on factors such as rate and storage requirements, but it must always include positive overflow from the system. Standing water must never be allowed in the porous pavement.

10. The subsurface bed and overflow should be designed and evaluated in the same manner as a detention basin to demonstrate the mitigation of peak flow rates. In this manner, the need for a detention basin will be eliminated or reduced significantly in size.

11. A weir plate or weir within an inlet or overflow control structure may be used to maximize the water level in the stone bed while allowing sufficient cover for overflow pipes.

12. Perforated pipes along the bottom of the bed may be used to distribute runoff evenly over the entire bed bottom. Continuously perforated pipes are connected to structures (such as cleanouts and inlet boxes). Pipes must lay flat along the bed bottom and provide for uniform distribution of water. Depending on size, these pipes may provide additional storage volume.

13. Roof leaders and area inlets may be connected to convey runoff water from adjacent areas to the bed. Water quality inserts or sump inlets are used to prevent the conveyance of sediment and debris into the bed.

14. Infiltration areas should be located within the immediate project area to control runoff at its source. Expected use and traffic demands are also considered

in pervious pavement placement. An impervious water stop should be placed along infiltration bed edges where pervious pavement meets standard impervious pavements.

15. Control of sediment is critical. Rigorous installation and maintenance of erosion and sediment control measures are required to prevent sediment deposition on the pavement surface or within the stone bed. Nonwoven geotextile may be folded over the edge of the pavement until the site is stabilized. The designer should consider the site placement of pervious pavement carefully to reduce the likelihood of sediment deposition. Surface sediment should be removed by a vacuum sweeper and not power-washed into the underlying bed.

16. Infiltration beds may be placed on a slope (Figure 8-1) by benching or terracing parking bays. Orienting parking bays along existing contours will reduce site disturbance and cut-and-fill requirements. If infiltration beds are to be terraced, earthen berms should be left in place between the various terraces to maximize storage and infiltration throughout the system.

17. The underlying infiltration bed is typically 12 to 36 in. deep and consists of clean, uniformly graded aggregate with approximately 40% void space (Figure 8-2). AASHTO No.3, which ranges from 1.5 to 2.5 in. in gradation, is often used. Depending on local aggregate availability, both larger and smaller aggregate has been used. The critical requirements are that the aggregate be uniformly graded, clean washed, and result in a significant void content. The depth of the bed is a function of stormwater storage requirements, site grading, and anticipated loading. Infiltration beds are typically sized to mitigate the increased runoff volume from the more frequent storm events.

18. While most pervious pavement installations are underlain by an aggregate bed, alternative subsurface storage products may also be employed. These

Figure 8-1 Infiltration bed system on a slope. (Courtesy of SKB Site Design.)

Figure 8-2 Example of stone aggregate for infiltration–storage bed. Uniformly graded 2 in. crushed with 40% void space.

Figure 8-3 Storage chambers fabricated from recycled plastics.

include a variety of proprietary, interlocking plastic units (Figure 8-3) that contain much greater storage capacity than aggregate, at an increased cost. Designers are encouraged to search the marketplace for alternative storage options.

19. All pervious pavement installations must have a backup method for water to enter the stone storage bed in the event that the pavement fails or has its permeability otherwise degraded. In uncurbed lots, this backup drainage may consist of an unpaved 2-ft-wide stone edge drain connected directly to the bed

Figure 8-4 Edge drain integral to bed, with riverstone surface.

(Figure 8-4). In curbed lots, inlets with sediment traps may be required at low spots. Backup drainage elements will ensure the functionality of the infiltration system if the permeability pervious pavement is ever compromised.

20. In areas with poorly draining soils, infiltration beds below pervious pavement may be designed to discharge slowly to adjacent wetlands or bioretention areas. In this way, a pervious pavement installation may act as an alternative form of capture and reuse for landscape irrigation. Only in extreme cases (e.g., industrial sites with contaminated soils) will the aggregate bed need to be lined to prevent infiltration.

21. In those areas where the threat of spills and groundwater contamination is likely, pretreatment systems such as filters and wetlands may be required before any infiltration occurs. In hot-spot areas such as truck stops and fueling stations, the appropriateness of pervious pavement must be considered carefully. A stone infiltration bed located beneath standard pavement, preceded by spill control and water quality treatment, may be more appropriate.

22. The use of pervious pavement must be considered carefully in areas where the pavement may be seal-coated or paved over due to lack of awareness, such as individual home driveways. In those situations, a system that is not easily altered by the property owner may be more appropriate. An example would include an infiltration system constructed under a conventional driveway. Educational signage at pervious pavement installations may guarantee its prolonged use in some areas.

Storage/Infiltration Bed Dimensions

The sizing of an infiltration bed is a function of the site layout and the opportunities presented. Unlike detention system designs, porous pavement design requires that careful thought be given to potential locations as part of the initial site layout, rather than as an afterthought based on the remaining vacant space after all other structures have been situated.

Infiltration Area The infiltration area is defined as the plan area of the storage reservoir under the pervious pavement. The minimum infiltration area should be no less than one-fifth the total contributing impervious surfaces, or a 5 : 1 ratio of impervious surface to infiltration bed area. The storage bed below the pavement provides runoff volume storage during storm events, while the undisturbed subgrade allows infiltration of runoff into the underlying soil mantle. The storage capacity of the stone bed can be calculated as follows:

$$\text{storage volume} = \text{depth(ft)} \times \text{area (ft}^2\text{)} \times \text{void space (e.g., 0.40 for aggregate)}$$

where depth is the depth of the water stored during a storm event, depending on the drainage area, conveyance to the bed, and outlet control elevation/configuration.

The amount of runoff that a given bed can infiltrate is as follows:

$$\text{infiltration volume} = \text{bed bottom area (ft}^2\text{)} \times \text{infiltration design rate (in./hr)} \times \text{infiltration period (hr)} \times 1/12$$

where infiltration period is the time when bed is receiving runoff and is capable of infiltrating at the design rate, not to exceed 72 hours.

Peak-Rate Mitigation Properly designed pervious pavement systems provide effective management of peak rates. The infiltration bed below the pavement acts as a storage reservoir during large storm events, even while runoff exfiltrates through the soil mantle through the process of infiltration. Outlet structures can be designed to manage peak rates with the use of weir and orifice controls, and carefully designed systems may be able to manage peak rates for storms up to and including the 100-year storm. For additional information relating to peak-rate modeling and routing, refer to Chapter 6.

Construction Staging

The sequence of steps required during construction of a porous pavement system is as follows:

1. Due to the nature of construction sites, pervious pavement and other infiltration measures should be installed toward the end of the construction period,

if possible. Infiltration beds under pervious pavement may be used as temporary sediment basins or traps provided that they are not excavated to within 12 in. of the designated bed bottom elevation. Once the site is stabilized and sediment storage is no longer required, the bed is excavated to final grade and the porous pavement system is installed.

2. The existing subgrade under the bed areas must not be compacted or subject to excessive construction equipment traffic prior to geotextile and stone bed placement.

3. Where erosion of subgrade has caused accumulation of fine materials and/or surface ponding, this material should be removed with light equipment and the underlying soils scarified to a minimum depth of 6 in. with a York rake (or equivalent) and light tractor. All fine grading is done by hand. All bed bottoms should be level grade.

4. Earthen berms (if used) between infiltration beds are left in place during excavation. These berms do not require compaction if proven stable during construction.

5. Geotextile and bed aggregate are placed immediately after approval of subgrade preparation (Figure 8-5). Geotextile is to be placed in accordance with manufacturer's standards and recommendations. Adjacent strips of geotextile should overlap a minimum of 18 in. It must also be secured at least 4 feet outside the bed to prevent any runoff or sediment from entering the storage bed. This edge strip should remain in place until all bare soils contiguous to beds have been stabilized and vegetated. Once the site is fully stabilized, excess geotextile along bed edges can be cut back to the bed edge.

6. Clean (washed) uniformly graded aggregate is placed in the bed in 8-in. lifts. Each layer should be lightly compacted, with the construction equipment

Figure 8-5 Placement of geotextile beneath infiltration beds, SMS site (now Siemens site), Chester County, Pennsylvania, 1981.

Figure 8-6 Permeability testing of porous pavement with water truck, Kaiser-Permanente Hospital, Modesto, California.

kept off the bed bottom as much as possible. Once bed aggregate is installed to the desired grade, a ±1-in. layer of choker base course (AASHTO 57 or equivalent) aggregate is installed uniformly over the surface to provide an even surface for paving.

7. Install pervious pavement. After final pervious asphalt or concrete installation, no vehicular traffic of any type is permitted on the pavement surface until cooling and hardening or curing has taken place, and in no case within the first 72 hours.

8. The full permeability of the pavement surface is tested by application of clean water at the rate of at least 5 gal/min over the surface using a hose or other distribution devise (Figure 8-6). All water applied must infiltrate directly without puddle formation or surface runoff.

Operation and Maintenance

While porous pavement does not require any type of "operation" in a typical engineering sense, it does warrant periodic inspection to assure continued performance. The primary goal of pervious pavement maintenance is to prevent the pavement surface and/or underlying infiltration bed from being clogged with fine sediments. To keep the system clean throughout the year, and to prolong its lifespan, the pavement surface should be vacuumed biannually using a commercial cleaning unit. All inlet structures within or draining to the infiltration beds should also be cleaned out on a biannual basis.

Planted areas adjacent to pervious pavement should be well maintained to prevent soil washout onto the pavement. If any washout does occur, it should be cleaned off the pavement immediately to prevent further clogging of the pores. Furthermore, if any bare spots or eroded areas are observed within the planted areas, they should be replanted and/or stabilized at once. Planted areas should be inspected on a semiannual basis. All trash and other litter that is observed during these inspections should be removed.

Superficial dirt does not necessarily clog the pavement voids, but dirt that is ground in repeatedly by tires can lead to clogging. Therefore, trucks or other heavy vehicles should be prevented from tracking or spilling dirt onto the pavement. Furthermore, all construction or hazardous materials carriers should be prohibited from entering a pervious pavement lot. Descriptive signage is recommended to maintain institutional memory of pervious pavement.

Vacuuming

Pervious pavement should be cleaned with a vacuum sweeper on a biannual basis. Acceptable types of vacuum sweepers include the Elgin Whirlwind and the Allianz Model 650. Although much less effective than "pure" vacuum sweepers, regenerative air sweepers (e.g., Tymco Model 210, Schwarze 348, Victory) are sometimes used. These units contain a blower system that generates a high-velocity air column, which forces the air against the pavement at an angle, creating a "peeling" or "knifing" effect. The high-volume air blast loosens the debris from the pavement surface, then transports it across the width of the sweeping head and lifts it into the containment hopper via a suction tube. Thus, sediment and debris are loosened from the pavement and sucked into the unit. Simple broom sweepers are not recommended for pervious pavement maintenance.

If the pavement surface has become clogged significantly, such that routine vacuum sweeping does not restore permeability, a more intensive level of treatment may be required. Recent studies have revealed the usefulness of washing pervious pavements with clean, low-pressure water, followed by immediate vacuuming. Combinations of washing and vacuuming techniques have proved effective in cleaning both organic and sandy clogging. Research in Florida found that a "power head cone nozzle" that "concentrated the water in a narrowly rotating cone" worked best. (*Note:* If the pressure of the washing nozzle is too great, contaminants may be driven farther into the pervious surface.) Maintenance crews are encouraged to determine the most effective strategy of cleaning their pervious pavement installations.

For smaller installations such as sidewalks, plazas, or small parking lots, walk-behind vacuum units may prove most effective. Although these units can be loud and somewhat deleterious to the operator due to the lack of dust suppression, they are also relatively easy to operate and inexpensive. Examples of acceptable walk-behind units, such as Billy Goat models, the Tennant 5700 industrial-strength scrubber, and sidewalk class vacuum sweepers (e.g., Nilfisk, Advance, Hako), are available. If walk-behind units are used, it is recommended that the scrub pressure be kept relatively low. The dirtiest areas may need to be power-washed after scrubbing to get out deeply ground-in dirt.

Restoration of Porous Pavements

Potholes in pervious pavement are extremely unlikely, although settling might occur if a soft spot in the subgrade is not removed during construction. For damaged areas of less than 50 square feet, a declivity could be patched by any means suitable with standard pavement, with the loss of porosity of that area being insignificant. The declivity can also be filled with pervious mix or paver units. If an area greater than 50 square feet is in need of repair, approval of the patch type must be sought from either the engineer or the owner. Under no circumstance is the pavement surface ever to be seal-coated. Any required repair of drainage structures should be done promptly to ensure continued proper functioning of the system.

With minimal maintenance, pervious pavement can function effectively for well over 20 years (the oldest system still in operation was installed in 1982 (see Figure 4-5). However, in the event that maintenance of the pervious pavement is neglected and it becomes clogged over time, the owner should vacuum the lot until the original permeability is restored. (If the original permeability of the lot cannot be restored, the pavement should be removed and replaced with a new pervious mix or pervious units.) Recent research has shown that one of the most effective ways of restoring pervious pavement is by applying a pressurized dose of a nontoxic detergent cleaning solution, allowing adequate soak time, and then vacuuming using a high-performance unit (e.g., Elgin Whirlwind, Allianz Model 650). Once again, it is important to note that high-pressure washing may drive contaminants farther into the pervious surface, even into the underlying aggregate. It is therefore recommended that prior to vacuum sweeping, a low-performance pressure washer be used to get the solution to break the surface tension and reach into the pores.

Pervious pavement maintenance considerations can be summarized as follows:

- The surface should never be seal-coated.
- Inspect for pavement rutting or raveling on an annual basis (some minor ruts may occur in pervious pavement from stationary wheel rotation).
- Damaged areas of less than 50 ft^2 can be patched with pervious or standard pavement.
- Larger areas should be patched with an approved pervious pavement.

Cost of Porous Pavement

The majority of added cost of a pervious pavement or infiltration system lies in the underlying stone bed, which is generally deeper than a conventional subbase and lined with nonwoven geotextile. However, for new construction projects, this additional cost is often offset by the significant reduction in the drainage infrastructure required (i.e., inlets and pipes). Also, since pervious pavement areas are often incorporated into the natural topography of a site, fewer earth-work and/or deep excavations are generally involved. Perhaps most significantly,

Table 8-1 Pavement Cost (per ft^2), San Diego County, California, 2005

	Demolition Cost	Subbase and Excavation	Pavement Installation	Total Cost	Annual Maintenance Cost
Porous AC, 18 in. backfill	$3.75	$1.88	$1.87	$6.50	$0.04
Standard AC, 6 in. backfill	$2.13	$1.04	$1.32	$4.49	$0.06
Porous concrete, 15 in. backfill	$3.19	$1.88	$6.34	$11.41	$0.02
Standard concrete, no backfill	$1.51	—	$3.42	$4.93	$0.01
Porous pavers, 18 in. backfill	$2.75	$1.88	$9.63	$14.26	TBD[a]

[a]TBD, to be determined.

pervious pavement areas with subsurface infiltration beds often eliminate the need (and associated costs, space, etc.) for large detention basins. When all of these factors are considered, pervious pavement with infiltration has proven itself less expensive than impervious pavement with associated traditional stormwater management. Pavement and stormwater management of recent installations has averaged between $2,000 and $2,500 per parking space.

Pervious asphalt, with additives, is generally 10 to 20% higher in cost than standard asphalt on a unit area basis. Unit costs for pervious asphalt (without an infiltration bed) range from about $1.75 to $3.50 per square foot. Pervious concrete as a material is generally more expensive than asphalt and requires more labor and experience for installation, due to specific material constraints. Unit costs for a 6-in.-thick pervious concrete (without an infiltration bed) section are typically between $6 and $7 per square foot. Pervious paver blocks vary in cost depending on type and manufacturer, and the data provided are based on average market costs. For greater accuracy, a site and market specific cost estimate should be developed.

Table 8-1 summarizes the costs associated with a pervious pavement demonstration project completed at the San Diego County Operations Center in Kearny Mesa. Phase I included pervious asphalt, concrete, and pavers, while phase II included only pervious asphalt (different mixes than in phase I) and concrete.

8.2 BIOREMEDIATION SYSTEMS

Rain Gardens

Design Considerations Rain gardens allow flexible design and can vary in complexity according to site conditions and runoff volume requirements. Although a rain garden is a structural BMP, the initial location of bioremediation areas should respect the site design process, and they should be integrated with preventive nonstructural BMPs. Design and construction procedures may vary from very simple for "backyard" rain gardens to highly engineered bioremediation cells in ultraurban areas.

It is important to note that rain gardens should not be confused with constructed wetlands or wet ponds that impond water permanently. Rain gardens are best suited for areas with at least moderate infiltration rates (more than 0.25 in./hr). In situations where permeability is less than 0.25 in./hr, special variations may apply, such as using amended soils or underdrains.

Rain gardens are often suitable for retrofit projects, through integration with previously developed lots and sites. An important concern for all rain garden applications is their long-term protection and maintenance, especially if undertaken in multiple (adjacent) residential lots where individual homeowners provide maintenance. In such situations it is important to provide management guarantees that ensure their long-term functionality, such as deed restrictions, covenants, and maintenance agreements.

1. *Sizing criteria*
   a. *Surface area* depends on storage volume requirements and permeability of the underlying native soil, but should generally not exceed a maximum loading ratio of 5 : 1 impervious drainage area to rain garden area. However, for design purposes, the total volume of water generated from the contributing drainage area must be used, not just the impervious portion.
   The rain garden surface area required is determined by taking the volume of runoff to be controlled according to LID criteria, maintaining the maximum ponding depth, the loading rate, and the emptying time. Infiltration and evapotranspiration are increased by increasing the surface area of the rain garden. The total surface area needed may be divided into multiple cells. This configuration, for example, may be useful to collect runoff from both the front and back of a building.
   b. *Surface side slopes* should be gradual. For most areas, maximum 3:1 side slopes are recommended; however, where space is limited, 2:1 side slopes may be acceptable.
   c. *Surface ponding depth* should not exceed 12 to 18 in. in most cases and should empty within 24 to 48 hours.
   d. *Ponding area* should provide sufficient surface area to meet required storage volume without exceeding the design ponding depth. The subsurface infiltration bed is used to supplement surface storage where appropriate.
   e. *Planting soil depth* should generally be between 24 and 36 in. where only herbaceous plant species will be utilized. If trees and woody shrubs will be used, soil media depth may need to be increased, depending on plant species (especially in poorly drained sites). Provided that they meet drainage criteria, native soils can be used as planting soil or can be modified on many sites.

2. *Planting soil* should be capable of supporting a healthy vegetative cover. Soils may need to be amended with a composted organic material. A recommended range of planting soil components is a mixture of 50 to 80% organic material (compost), 20 to 40% sand, and 0 to 10% topsoil, although any soil with sufficient drainage may be considered for a rain garden. Planting soil should be approximately 4 in. deeper than the bottom of the largest root ball.

3. *Existing soils* should also have a pH between 5.5 and 6.5 (better pollutant adsorption and microbial activity), a clay content of less than 10% (a small amount of clay is beneficial to adsorb pollutants and retain water, although no clay is necessary if pollutant loadings are not an issue), be free of toxic substances and unwanted plant material, and have a 5 to 10% organic matter content. Additional organic matter can be added to the soil to increase the water-holding capacity. If the void space within an amended soil mix will be used in the calculation of runoff volume capacity in the system, tests should be conducted on the soil's porosity to determine the storage capacity available.

4. To be effective, proper *plant selection* is essential for bioremediation areas. Typically, generalist plant species native to the area are best suited to the variable environmental conditions encountered in a rain garden, as they need to withstand a wide range of soil and moisture regimes. Recommended plant lists of suitable species are available in all parts of the country. When designing the planting, it is important that plant species are located according to their tolerance of inundation and prolonged saturated soils; less tolerant species should be located at the higher elevations of a rain garden. Trees, shrubs, and herbaceous perennials all may be used in a rain garden; they should be selected with other functions in mind (e.g., shade, screening versus clear views, color) in addition to suitability for bioremediation. For rain gardens that will have an underdrain, it is also important to select species that do not have invasive roots, which have a tendency to clog perforated drainage pipes. A landscape architect should help with plant selection and rain garden design.

5. *Planting time* should occur during fall to take advantage of seasonal rains in most regions. In most cases, seed is not the preferred method for plant establishment in a rain garden. The fluctuating water levels make it difficult for the seed to establish readily, and the random nature of seeding may result in an undesirable plant layout for some situations. Instead, it is strongly recommended that containerized live plants be utilized: plugs or 1-gallon container for herbaceous plants, 1- to 5-gallon container for shrubs, and 5-gallon container to 24-in. box for trees. Plant spacing depends on mature plant size and the desired density of plant cover.

6. A maximum of 2 to 3 in. of shredded hardwood *mulch* (aged at least 6 months to 1 year), leaf compost, or comparable product should be applied uniformly immediately after planting to prevent erosion, enhance metal removal, and aid plant establishment. Wood chips should be avoided, as they tend to float during inundation periods. Mulch or compost should not exceed 3 in. in depth so

as not to restrict oxygen flow, or be placed directly against the stems or trunks of plants.

7. When working in areas with *steeper slopes* (up to 15%), it is critical first to ensure that these BMPs are feasible. A geotechnical engineer should be consulted to evaluate the suitability of installing a rain garden on or near a steep slope to identify the risk of creating an unstable condition; underdrains may be required for slope applications. When they do occur on slopes, rain gardens should be terraced laterally along slope contours to minimize earthwork and provide level areas for infiltration.

8. *Underdrains* are generally not needed unless in situ soils are expected to cause ponding lasting longer than 48 hours (e.g., type C and D soils), for slopes where soil instability is likely, and where there is insufficient space (<10 ft) between the rain garden and a structure. Because infiltration is encouraged, detailed discussion of underdrain design is not included in this book but can be reviewed in other guidance documents.

9. *Surface infiltration area* is the average area of a rain garden, defined as

$$A_{\text{inf}} = \frac{\text{area of rain garden at ponding depth} + \text{bottom area of rain garden}}{2}$$

$$= \text{infiltration area (average area)}$$

The size of the infiltration area necessary is determined by the volume of water necessary to remove as determined by LID criteria, the depth of the ponded area (not to exceed 1 ft), the infiltration rate of the soil, the loading ratio, and if applicable, any subsurface storage in the amended soil or gravel.

Storage Volume The storage volume of a rain garden is defined as the sum of the surface and subsurface void volumes beneath the level of the discharge invert. Intermedia void volumes may vary considerably based on design variations. Void volumes should only be considered in soils when they have been amended to a known depth with a known void ratio. If the rain garden is placed in existing natural soils, the infiltration rate may be used to calculate discharge out of the basin into the ground and the additional capacity available as a result.

The storage volume of a rain garden has three components:

1. Surface storage volume (ft^3) = average bed area (ft^2) × average design water depth
2. Soil storage volume (ft^3) = infiltration bed area (ft^2) × depth of amended soil (ft) × void ratio of amended soil
3. Subsurface storage/infiltration bed volume (ft^3) = infiltration area × void ratio of storage material

$$\text{Total rain garden volume} = \text{surface storage plus soil storage volume} + \text{subsurface storage (if applicable)}$$

Water Quality Mitigation The water quality benefits reported for rain gardens can be expected to remove a high amount of total suspended solids (typically, 70 to 90%), a medium amount of total phosphorus (approximately 60%), and a medium amount of total nitrogen (often 40 to 50%). In areas with high sediment loading, pretreatment of runoff can significantly reduce the amount of rain garden maintenance required. See Chapter 6 for water quality compliance procedures.

Construction of a Rain Garden

The following is a typical construction sequence. However, alterations will be necessary depending on design variations.

1. Install temporary sediment control BMPs as required by permitting authority.
2. Complete the site grading, minimizing compaction as much as possible. If applicable, construct curb cuts or other inflow entrance but provide protection so that drainage is prohibited from entering the rain garden construction area. Construct pretreatment devices (e.g., filter strips, swales) if applicable.
3. Stabilize grading except within the rain garden area. Rain garden bed areas may be used as temporary sediment traps provided that the proposed finish elevation of the bed is at least 12 in. lower than the bottom elevation of the sediment trap.
4. Excavate the rain garden to proposed invert depth, and scarify the existing soil surfaces. Do not compact soils.
5. If applicable, install slotted underdrain and/or gravel drainage or storage layer.
6. Backfill the rain garden with amended soil as shown on plans and specifications. Overfilling is recommended to account for settling. Light hand tamping is acceptable if necessary.
7. Install an automatic irrigation system if applicable.
8. Prewet the planting soil at least 24 hours before planting to aid in settlement.
9. Complete final grading to achieve proposed design elevations, leaving space for the upper layer of compost or mulch as specified on plans.
10. Plant vegetation according to the planting plan.
11. Apply the mulch layer.
12. Install erosion protection at surface flow entrances where necessary.

Maintenance of Rain Gardens

Rain gardens designed and installed properly require some regular maintenance, most frequently during the first year or two of establishment.

1. Rain gardens will require supplemental irrigation during the first 2 to 3 years after planting. Drought-tolerant species may need little additional water

after this period, except during prolonged drought or if a "fresher" appearance is required. Ensure that the maintenance plan includes a watering schedule for the establishment period and in times of extreme drought after plants have been established.

2. While vegetation is being established, remove any weeds by hand (weeding frequency should decrease over time, as rain garden plants grow).

3. Although plants may need occasional pruning or trimming, rain gardens should not be mowed on a regular basis. Trim vegetation as necessary to maintain healthy plant growth.

4. Replace dead plants. If a particular species is more prone to mortality, it may need to be replaced with a different species that is more likely to succeed in the rain garden.

5. Mulch should be reapplied when erosion is evident. In areas expected to have low metal loads in the runoff, mulch as needed to maintain a depth of 2 to 3 in. In areas with relatively high metal loads, replace the mulch once per year.

6. Rain gardens should be inspected at least twice a year for sediment buildup, erosion, and to evaluate the health of the vegetation. If sediment buildup reaches 25% of the ponding depth, it should be removed. If erosion is noticed within the rain garden, additional soil stabilization measures should be employed. If vegetation appears to be in poor health with no obvious cause, a landscape specialist should be consulted.

Cost of Rain Gardens

Rain gardens often replace areas that would have been landscaped and maintenance-intensive, so the net cost can be less than the conventional alternative. In addition, the use of rain gardens can decrease the cost for stormwater conveyance systems on a site. Rain gardens cost approximately $5 to $7 per cubic foot of storage to construct.

Vegetated Roof Systems

Design Considerations

1. The live- and dead-load bearing capacity of the roof must be established. Dead loads should be estimated using media weights determined using a standardized laboratory procedure.

2. Waterproofing materials must be durable under the conditions associated with vegetated covers. A supplemental root-barrier layer should be installed in conjunction with materials that are not root-fast.

3. Roof flashings should extend 6 in. higher than the top of the growth medium surface and be protected by counter flashings.

4. The design should incorporate measures to protect the waterproofing membrane from physical damage during and after installation of the vegetated cover assembly.

5. Vegetated roof covers should incorporate internal drainage capacity sufficient to accommodate up to a 2-year-return-frequency rainfall without generating surface runoff flow.

6. Deck drains, scuppers, or gravel stops serving as methods to discharge water from the roof area should be protected with access chambers. These enclosures should include removable lids to allow ready access for inspection.

7. The design and layout should provide for maintenance access at least around the roof perimeter and to other rooftop components (i.e., vents, HVAC equipment, etc.).

8. The physical properties of the engineered media should be selected to achieve the desired hydrologic performance and meet weight limitations.

9. Engineered media should contain no clay-size particles and should contain no more than 15% organic matter (wet combustion or loss on ignition methods).

10. Media used in constructing vegetated roof covers should have a maximum moisture capacity between 30 and 40%.

11. The irrigation system, if provided, should preferably consist of subsurface drip or capillary mat water delivery, to minimize water loss by evaporation.

12. Plants should be selected that will create a healthy, drought-tolerant roof cover (Figures 8-7 and 8-8). In general, several criteria should help guide species selection for extensive vegetated roofs:

a. Native or adapted species tolerant of extreme climate conditions (e.g., heat, drought, wind; cold for high-elevation and far inland areas): a "generalist."

b. Low-growing, with a range of growth forms: spreading evergreen shrubs or subshrubs, succulents, perennials, self-seeding annuals; note

Figure 8-7 Regency House, Washington, DC. (Courtesy of Roofmeadow, Inc.)

Figure 8-8 Boston World Trade Center, Boston, Massachusetts. (Courtesy of Roofmeadow, Inc.)

that bunchgrasses and resinous shrubs may not be approved for projects that must follow strict fire fuel management guidelines.

c. A variety of species and growth forms should be considered for a vegetated roof project, to ensure survival and growth of species that happen to succeed for the project's particular conditions.

d. Because many perennials and annuals are dormant during part or all of the rainy season, evergreen and cool-season plants should be included to help with rainfall interception and evapotranspiration when rains typically occur.

e. Plant species should not develop a deep taproot.

f. Long life (or self-propagating), with low maintenance and fertilizer needs.

g. Mature plant size should be factored into the load-bearing capacity of the roof (e.g., mature perennials can add approximately 2 to 5 lb/ft^2).

13. Roofs with pitches exceeding 2 : 12 (9.5°) will require supplemental measures to ensure stability against sliding.

Runoff Water Quality Mitigation Direct runoff from roofs is a contributor to non-point source pollutant releases. Vegetated roof covers can significantly reduce this source of pollution. Assemblies intended to produce water quality benefits will employ engineered media with almost 100% mineral content. Furthermore, following the plant establishment period (usually about 18 months), ongoing fertilization of the cover should not be permitted, unless plants exhibit signs of nutrient stress (then a low-strength organic fertilizer could be applied annually). Experience indicates that it may take 5 or more years for a water-quality vegetated cover to attain its maximum pollutant removal efficiency.

In Combination with Infiltration Measures Vegetated roof covers are frequently combined with ground infiltration measures. Vegetated roofs improve the efficiency of infiltration devices by:

- Reducing the peak runoff rate
- Prolonging the runoff
- Filtering runoff to produce a cleaner effluent

Roofs that are designed to achieve water quality improvements will also reduce pollutant inputs to infiltration devices.

Construction of a Vegetated Roof

Inspect the completed waterproofing visually to identify any apparent flaws, irregularities, or conditions that will interfere with the security or functionality of the vegetated cover system. The waterproofing should be tested for watertightness by the roofing applicator.

1. Institute a program to safely install the vegetated roof system.
2. Introduce measures to protect the finished waterproofing from physical damage.
3. Install slope-stabilization measures (pitched roofs with pitches in excess of 2 : 12).
4. If the waterproofing materials are not root-fast, install a root-barrier layer.
5. Lay out key drainage and irrigation components, including drain access chambers, internal drainage conduit, confinement border units, and isolation frames (for rooftop utilities, hatches, and penetrations).
6. Install walkways and paths (for maintenance or projects with public access).
7. Test irrigation systems (as relevant for roof gardens).
8. Install the drainage layer. Depending on the variation type, this could be a geocomposite drain, mat, or drainage medium.
9. Provide protection (e.g., ultraviolet-degradable erosion control netting) from wind disruptions as warranted by the project conditions and plant establishment method.

Note: In some installations, slope-stabilizing measures can be introduced as part of the roof structure and will be already be in place at the start of the construction sequence.

Maintenance of Vegetated Roofs

1. During the plant establishment period (approximately 2 years), periodic irrigation will be required. After establishment, irrigation may be needed during dry periods.

2. During the plant establishment period, visits every 3 to 4 months are recommended to conduct basic weeding, fertilization (if needed), and infill planting. Thereafter, only two visits per year for inspection and light weeding should be required (irrigated assemblies will require more intensive maintenance).

3. Drainage outlets should be inspected periodically to ensure that they drain freely and are not clogged with debris.

4. The roofing membrane should be inspected periodically for drainage or leaks. Leaks may be pinpointed using electric field vector mapping, provided that the vegetated roof assembly and roof deck material are amenable to this technique.

Cost of Vegetated Roofs

The construction cost of vegetated roof covers can vary greatly, depending on factors such as:

- Height of the building
- Accessibility to the structure by large equipment such as cranes and trailers
- Depth and complexity of the assembly
- Remoteness of the project from sources of material supply
- Size of the project

However, under present market conditions (2011), extensive vegetated roofs will typically range between $13 and $18 per square foot, including design, installation, and warranty service (not including waterproofing, which can potentially double the cost). Basic maintenance for extensive vegetated covers typically requires about 3 person-hours per 1,000 ft^2, annually.

8.3 CAPTURE–REUSE SYSTEMS

Construction

1. Calculate the annual yield from the proposed rooftop area, assuming 95% capture efficiency. Then estimate building demand on both an annual and a seasonal basis, including external use for irrigation. If demand exceeds potential supply, the system must be coupled with the public supply system. If potential roof yield is sufficient for irrigation needs, the capture system can be independent but the storage capacity must equal the demand storage. In southern California, the end of the rainy season and the beginning of the irrigation season is about

4 months, so the storage capacity must equal that demand. Again, if potential capture is insufficient for total seasonal needs, public supply must augment the system.

2. Where potable use is anticipated, the storage may be coupled with public service, following guidelines established by purveyor and local health department regulations. Filtration and disinfection units must be approved by all regulatory agencies.

3. Some designs may identify opportunities where treated wastewater can be reused for irrigation. It is recommended that such systems be isolated from rooftop capture systems and stored separately, because of the potential for contamination of potable supply.

4. The estimate of irrigation water demand will be based on the selection of landscape materials. As discussed elsewhere in the book, a number of possible plant mixes can be designed that minimize the need for irrigation. Every effort should be made to utilize the guidance shown to reduce this demand as much as possible, assuming that in the near future, the use of potable water for this purpose will be restricted. Based on the acreage in landscape and the minimum use, the designer must determine how much water will be needed to achieve this goal and how often the storage unit will be refilled via precipitation.

5. For residential design, rain barrels and cisterns should be positioned to receive rooftop runoff directly, especially with roof designs that have multiple panels. If only irrigation use is anticipated, the site design should be configured so that the planted beds are in close proximity to the downspouts. For potable use, single tank storage is most efficient for filtration, disinfection, and pressurization. The design capacity should probably be at least 2,000 gallons per dwelling unit, and greater if possible. External storage below grade may be most appealing from an aesthetic consideration.

6. Provide for the discharge or release of unconsumed stored water between storm events so that the necessary stormwater storage volume will be available and fully mitigate any runoff rate increase. Any release should, where possible, be directed to an infiltration system.

7. Consider household water demands when sizing a system to supplement residential gray water use. Every effort should be made to minimize potable water usage, including low or ultralow flow fixtures and other techniques that reduce demand.

8. Discharge points or storage units should be clearly labeled "Caution: Untreated Rainwater: ***Do Not Drink***" unless an approved disinfection unit is installed in line with the service. In some countries, such as New Zealand, rooftop runoff is used for all residential demands following ultraviolet disinfection (Figure 8-9). It is appropriate that this example should be the last image in this book, because it represents the greatest potential concept in stormwater management: reuse for potable purposes.

9. Screens should be used to prevent roof debris from entering storage units.

Figure 8-9 New Zealand rooftop rainfall capture–reuse system.

10. Protect semitransparent storage elements from direct sunlight by positioning and landscaping. Limit light into devices to minimize algae growth.

11. Any storage unit overflow should be well removed from building foundations.

12. Cisterns should be watertight (joints sealed with nontoxic waterproof material) with a smooth interior surface.

13. Covers and lids should have a tight fit to keep out surface water, animals, dust, and light.

14. Positive outlet for overflow should be provided a few inches from the top of the cistern and should be sized to safely discharge the appropriate design storms when the cistern is full.

15. Observation risers should be at least 6 inches above grade for buried cisterns.

16. Reuse may require pressurization. Water stored has a pressure of 0.43 psi per foot of water elevation. A 10-foot tank would have a head of $0.43 \times 10 = 4.3$ psi. Most irrigation systems require at least 15 psi, and building distribution systems operate in the range 30 to 50 psi. To add pressure, a pump and pressure tank can be used.

17. For potable use systems, disinfection in the form of ultraviolet radiation units, usually designed as linear flow-through tubes, is usually most practical and most efficient. These elements add to the cost of a system but protect users from any potential health threat or contamination.

18. Simple rain barrels require a release mechanism that allows them to empty between storm events. Connect a soaker hose to slowly release stored water to a landscaped area.

19. Capture–reuse can be achieved using a subsurface storage reservoir that provides temporary storage of stormwater runoff. The stormwater storage reservoir may consist of clean uniformly graded aggregate or premanufactured structural stormwater storage units. This type of system is very similar to a subsurface infiltration system except that infiltration is prevented with a waterproof liner or the presence of a restrictive soil layer, and a way to reuse the water must be provided (e.g., pump well, manhole, sump).

Volume Reduction

Based on the area of the roof structure, the potential annual yield can be estimated. The seasonal variability and demand cycle(s) must be developed to optimize the storage capacity required. After the water need is determined, choose a structure that is large enough to contain the amount of water needed. The amount of storage in the container is equal to the volume reduction.

Peak-Rate Mitigation

Overall, capture–reuse takes a volume of water out of site runoff and puts it into storage, effectively removing the volume of the rooftop rainfall from the site runoff. This reduction in volume will translate into a lower overall peak rate for the site. The proportional reduction is dependent on the size of the rooftop as compared to the total site impervious cover. In many sites, the use of this system will meet the regulatory requirements for volume reduction. However, the potential for pollutant production from the surrounding site landscape and impervious surfaces should not be neglected and may well warrant additional water quality mitigation measures, such the use of porous pavements and rain gardens.

Water Quality Mitigation

The water quality of rooftop runoff will vary with location, type and age of roofing material, and proximity of localized air pollutant emissions. Pollutant removal takes place through the settling of particulate matter in the storage unit and subsequent filtration and disinfection prior to consumption in the case of potable use. For nonpotable uses, filtration or disinfection is usually not required. In most locations, roof runoff is as clean or cleaner than surface waters and should serve well as a source of raw water.

APPENDIX A

THE STORMWATER CALCULATION PROCESS

Flowchart A is provided to guide the user through the first steps of the stormwater calculation process and can be thought of as a series of steps executed through a series of worksheets.

Step 1. Provide general site information (Worksheet 1).

Step 2. Identify sensitive natural resources, and if applicable, identify which areas will be protected (Worksheet 2).

Step 3. Identify the area of the site contributing to a capture–reuse system (Worksheet 3). Identify the area of the site for which stormwater runoff must be managed, or the stormwater management area (Worksheet 3). (*Note:* Steps 10 and 11 discuss the procedure for taking stormwater credit for capture–reuse strategies.)

Step 4. Incorporate nonstructural best management practices (BMPs) into the stormwater design. Quantify the benefits of nonstructural BMPs (Worksheet 3). For nonstructural BMPs being proposed, provide nonstructural BMP checklists to demonstrate that these BMPs are appropriate.

Step 5. Estimate the target volume of runoff associated with the design storm event (e.g., the 85th percentile storm runoff volume, the net increase in a 2-year storm), using the runoff curve number or small storm hydrology method. The use of weighted curve numbers for pervious and impervious areas is *not* acceptable. Runoff volume should be calculated based on major land use types and soil types (Worksheet 4).

Low Impact Development and Sustainable Stormwater Management, First Edition. Thomas H. Cahill.

Step 6. Design and incorporate structural BMPs that provide volume control for the design storm volume increase indicated on Worksheet 4. Provide calculations and documentation to support the volume estimate provided by BMPs. Indicate the permanent volume reduction provided by the BMPs proposed (Worksheet 5). Also, indicate the area of non-volume-reducing BMPs included in the design. BMPs that reduce volume include infiltration systems, bioretention, and pervious pavement and infiltration beds, among others (capture–reuse is accounted for separately in Worksheets 9 and 10). Proceed to Flowchart B.

Step 7. Determine applicable peak rate requirements and calculations. If applying for an exemption for small sites and small impervious area sites, complete Worksheet 6.

Step 8. If additional peak rate calculations are required, demonstrate peak rate mitigation through modeling. This modeling or routing should consider the benefits of volume-control BMPs as outlined in Flowchart B and below. Provide additional detention capacity if needed.

8.1. List the applicable design criteria (local requirement, LID guidance, or other) and what they requires on Worksheet 7.

8.2. Use one of the following methods to determine peak rate control for all storms up to the 100-year storm or as required by local requirements.

8.3. List the pre- and postdevelopment peak rates for each design storm in the space provided on Worksheet 7.

Other methods recommended for determining the effects of volume control on peak rate mitigation are listed in the next section. Proceed to Flowchart C.

Step 9. If the net increase in runoff volume associated with a 2-year, 24-hour storm (pre- vs. postdevelopment) is being controlled adequately on-site (i.e., reduced by infiltration and/or evapotranspiration), there is no need to complete Worksheet 8. However, if this net increase is not being controlled (i.e., permanently reduced), complete Worksheet 8. Select BMPs that will remove the expected pollutants for the land use type. Often, multiple types of BMPs used in series will be required to provide adequate treatment. As a guide, use a series of BMPs that will achieve 80% removal of solids or better.

Step 10. Determine the landscape irrigation demand (complete Worksheet 9). The irrigation demand is evaluated for each landscape type on a particular site by month. Worksheet 9 provides space for two types of landscape. If additional landscapes are proposed, the designer should use as many worksheets as necessary. Once all the landscape irrigation demands have been determined, the designer calculates the total landscape irrigation demand for the site by summing the individual demands by month.

Step 11. Determine the percentage of site area completely controlled (i.e., no runoff in an average year) by a capture–reuse system (capture–reuse area) that can be excluded from the stormwater management area (complete the process outlined on Worksheet 10). Starting with a design storage volume and contributing area, the designer can use Worksheet 10 to estimate the total runoff volume from a site (supply for capture–reuse system), the total overflow volume from the site (exceeding the design storage volume), and thus the total runoff volume captured and the percentage of runoff captured. The designer then applies the percentage of runoff captured to the contributing area to calculate the capture–reuse area, which can be subtracted from the total site area. Return to step 3 and complete Worksheet 3.

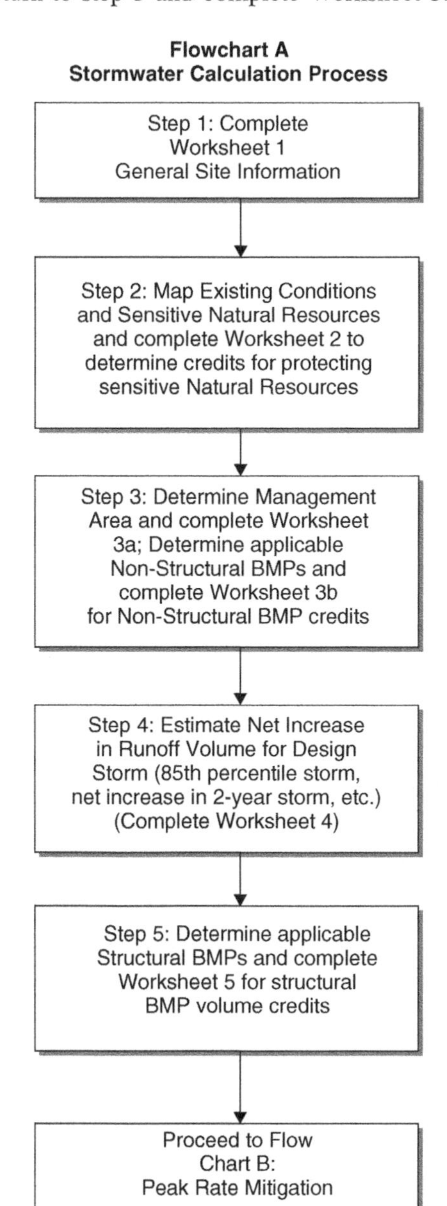

Worksheet 1. General Site Information

INSTRUCTIONS: Fill out Worksheet 1 for each watershed

Date: ______________________

Project Name: ______________________

Municipality: ______________________

County: ______________________

Total Area (acres): ______________________

Major River Basin: ______________________

http://www.dep.state.pa.us/dep/deputate/watermgt/wc/default.htm#newtopics

Watershed: ______________________

Sub-Basin: ______________________

Nearest Surface Water(s) to Receive Runoff: ______________________

Chapter 93 - Designated Water Use: ______________________

http://www.pacode.com/secure/data/025/chapter93/chap93toc.html

Impaired according to Chapter 303 (d) List? Yes ☐ No ☐

http://www.dep.state.pa.us/dep/deputate/watermgt/wqp/wqstandards/303d-Report.htm

List Causes of Impairment:

Is project subject to, or part of:

Municipal Separate Storm Sewer System (MS4) Requirements? Yes ☐ No ☐

http://www.dep.state.pa.us/dep/deputate/watermgt/wc/Subjects/StormwaterManagement/GeneralPermits/default.htm

Existing or planned drinking water supply? Yes ☐ No ☐

If yes, distance from proposed discharge (miles): ______________

Approved Act 167 Plan? Yes ☐ No ☐

http://www.dep.state.pa.us/dep/deputate/watermgt/wc/Subjects/StormwaterManagement/Approved_1.html

Existing River Conservation Plan? Yes ☐ No ☐

http://www.dcnr.state.pa.us/brc/rivers/riversconservation/pl/stcejorpgninna

Worksheet 2. Sensitive Natural Resources

INSTRUCTIONS:

1. Provide Sensitive Resources Map according to non-structural BMP in Chapter 4. This map should identify wetlands, woodlands, natural drainage ways, steep slopes, and other sensitive natural areas.

2. Summarize the existing extent of each sensitive resource in the Existing Sensitive Resources Table (below, using Acres).

3. Summarize Total Protected Area as defined under BMPs in Chapter 4.

4. Do not count any area twice. For example, an area that is both a floodplain and a wetland may only be considered once.

EXISTING NATURAL SENSITIVE RESOURCE	MAPPED? yes/no/n/a	TOTAL AREA (Ac.)	PROTECTED AREA (Ac.)
Waterbodies			
Floodplains			
Riparian Areas			
Wetlands			
Woodlands			
Natural Drainage Ways			
Steep Slopes, 15% – 25%			
Steep Slopes, over 25%			
Other:			
Other:			
TOTAL EXISTING:			

Worksheet 3. Nonstructural BMP Credits

PROTECTED AREA

1.1 Area of Protected Sensitive/Special Value Features (see WS 2) _____ Ac.

1.2 Area of Riparian Forest Buffer Protection _____ Ac.

3.1 Area of Minimum Disturbance/Reduced Grading _____ Ac.

TOTAL _____ Ac.

Site Area	minus	Protected Area	=	Stormwater Management Area
[ ]	−	[ ]	=	[ ]

This is the area that requires stormwater management

VOLUME CREDITS

3.1 Minimum Soil Compaction

Lawn _____ ft^2 x 1/4"x 1/12 = _____ ft^3

Meadow _____ ft^2 x 1/3"x 1/12 = _____ ft^3

3.3 Protect Existing Trees

For Trees within 100 feet of impervious area:

Tree Canopy _____ ft^2 x 1/2"x 1/12 = _____ ft^3

For Trees within 20 feet of impervious area:

Tree Canopy _____ ft^2 x 1"x 1/12 = _____ ft^3

5.1 Disconnect Roof Leaders to Vegetated Areas

For runoff directed to areas protected under 5.8.1 and 5.8.2

Roof Area _____ ft^2 x 1/3"x 1/12 = _____ ft^3

For all other disconnected roof areas

Roof Area _____ ft^2 x 1/4"x 1/12 = _____ ft^3

5.2 Disconnect Non-Roof impervious to Vegetated Areas

For Runoff directed to areas protected under 5.8.1 and 5.8.2

Impervious Area _____ ft^2 x 1/3"x 1/12 = _____ ft^3

For all other disconnected roof areas

Impervious Area _____ ft^2 x 1/4"x 1/12 = _____ ft^3

TOTAL NON-STRUCTURAL VOLUME CREDIT* [ ] ft^3

** For use on Worksheet 5*

WORKSHEET 4 . CHANGE IN RUNOFF VOLUME FOR 2-YR STORM EVENT

PROJECT: ____________________

DA: ____________________

2-Year Rainfall: ________ in

Total Site Area: ________ acres

Protected Site Area: ________ acres

Managed Area: ________ acres

Existing Conditions:

Cover Type/Condition	Soil Type	Area (sf)	Area (ac)	CN	S	Ia (0.2*S)	Q Runoff[1] (in)	Runoff Volume[2] (ft3)
Woodland								
Meadow								
Impervious								
TOTAL:								

Developed Conditions:

Cover Type/Condition	Soil Type	Area (sf)	Area (ac)	CN	S	Ia (0.2*S)	Q Runoff[1] (in)	Runoff Volume[2] (ft3)
TOTAL:								

2-Year Volume Increase (ft3):

2-Year Volume Increase = Developed Conditions Runoff Volume - Existing Conditions Runoff Volume

1. Runoff (in) = Q = $(P - 0.2S)^2 / (P + 0.8S)$ where

    P = 2-Year Rainfall (in)

    S = (1000/ CN)-10

2. Runoff Volume (CF) = Q x Area x 1/12

    Q = Runoff (in)

    Area = Land use area (sq. ft)

Note: Runoff Volume must be calculated for EACH land use type/condition and HSGI.
The use of a weighted CN value for volume calculations is not acceptable.

WORKSHEET 5 . STRUCTURAL BMP VOLUME CREDITS

PROJECT: ______________________

SUB-BASIN: ______________________

Required Control Volume (ft^3) - *from Worksheet 4:* __________

Non-structural Volume Credit (ft^3) - *from Worksheet 3:* __________

Structural Volume Reqmt (ft^3) __________
(*Required Control Volume minus Non-structural Credit*)

Proposed BMP*	Area (ft^2)	Storage Volume (ft^3)
6.4.1 Porous Pavement		
6.4.2 Infiltration Basin		
6.4.3 Infiltration Bed		
6.4.4 Infiltration Trench		
6.4.5 Rain Garden/Bioretention		
6.4.6 Dry Well / Seepage Pit		
6.4.7 Constructed Filter		
6.4.8 Vegetated Swale		
6.4.9 Vegetated Filter Strip		
6.4.10 Berm		
6.5.1 Vegetated Roof		
6.5.2 Capture and Re-use		
6.6.1 Constructed Wetlands		
6.6.2 Wet Pond / Retention Basin		
6.6.3 Dry Extended Detention Basin		
6.6.4 Water Quality Filters		
6.7.1 Riparian Buffer Restoration		
6.7.2 Landscape Restoration / Reforestation		
6.7.3 Soil Amendment		
6.8.1 Level Spreader		
6.8.2 Special Storage Areas		
Other		

Total Structural Volume (ft^3): __________

Structural Volume Requirement (ft^3): __________

DIFFERENCE __________

Flowchart B
Control Guideline 1 Process

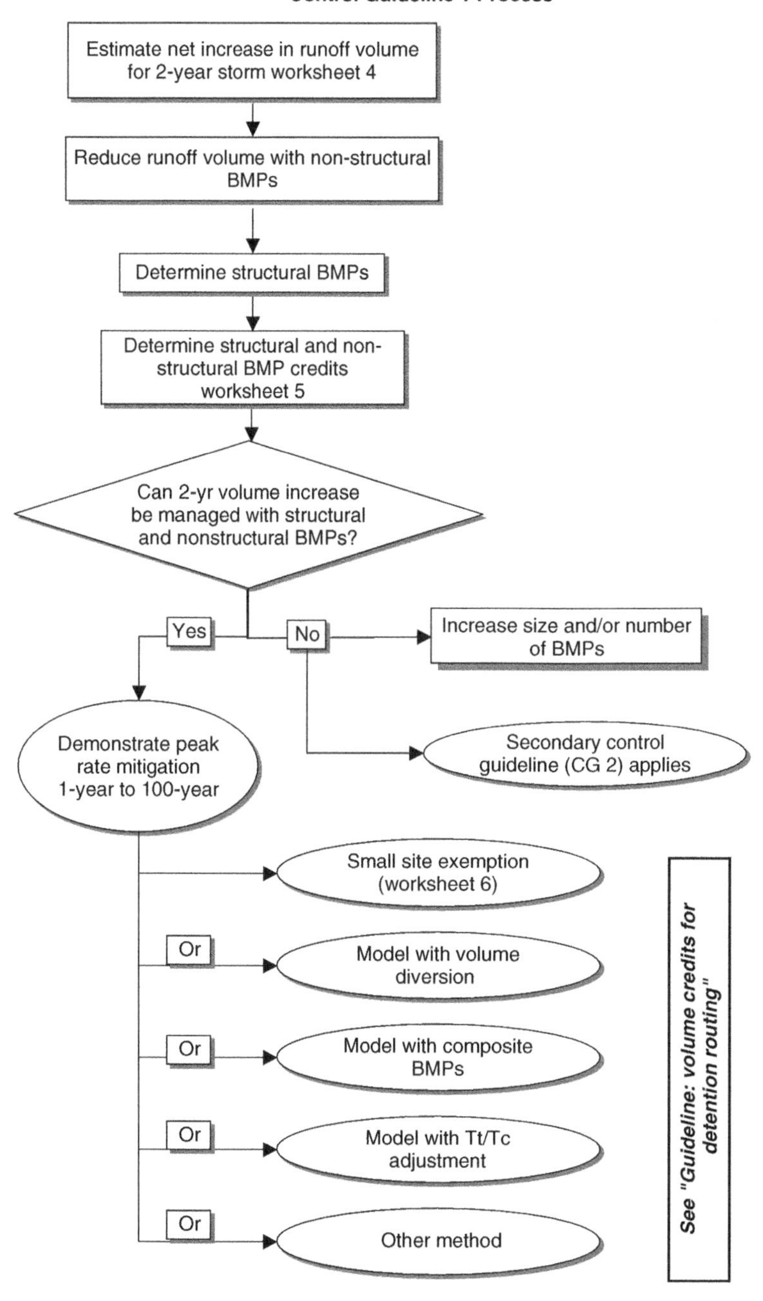

WORKSHEET 6 . SMALL SITE / SMALL IMPERVIOUS AREA EXCEPTION FOR PEAK RATE MITIGATION CALCULATIONS

The following conditions must be met for exemption from peak rate analysis for small sites under CG-1:

______ The 2-Year Runoff Volume increase must be met in BMPs designed in accordance with Manual Standards

______ Total Site Impervious Area may not exceed **1 acre.**

______ Maximum Development Area is **10 acres.**

______ Maximum site impervious cover is 50%.

______ No more than 25% Volume Control can be in Non-structural BMPs

______ Infiltration BMPs must have an infiltration rate of 0.5 in/hr.

Site Area	Percent Impervious	Total Impervious
10 acre	10%	1 acre
5 acre	20%	1 acre
2 acre	50%	1 acre
1 acre	50%	0.5 acre
0.5 acre	50%	0.25 acre

WORKSHEET 7. CALCULATION OF RUNOFF VOLUMES (PRV and EDV) FOR CG-2 ONLY

PROJECT: ____________________
DRAINAGE AREA:

Total Site Area: __________ acres
Protected Site Area: __________ acres
Managed Area: __________ acres
Total Impervious Area acres

2 Inch Runoff - Multiply Total Impervious Area by 2 inch

Cover Type	Area (ac)	Runoff Capture Volume (ft^3)
Roof		
Pavement		
Other Impervious		
TOTAL:		

1 Inch Rainfall -

Cover Type	Area (sf)	Area (ac)	Runoff (in)	Runoff Volumes (ft^3)
TOTAL:				

1. Total Runoff Capture Volume (ft^3) = Total Impervious Area (ft^2) x 2 inch x 1/12
2. PRV (ft^3) = Total Impervious Area (ft^2) x 1 inch x 1/12
3. EDV (ft^3) = Total Impervious Area (ft^2) x 1 inch x 1/12

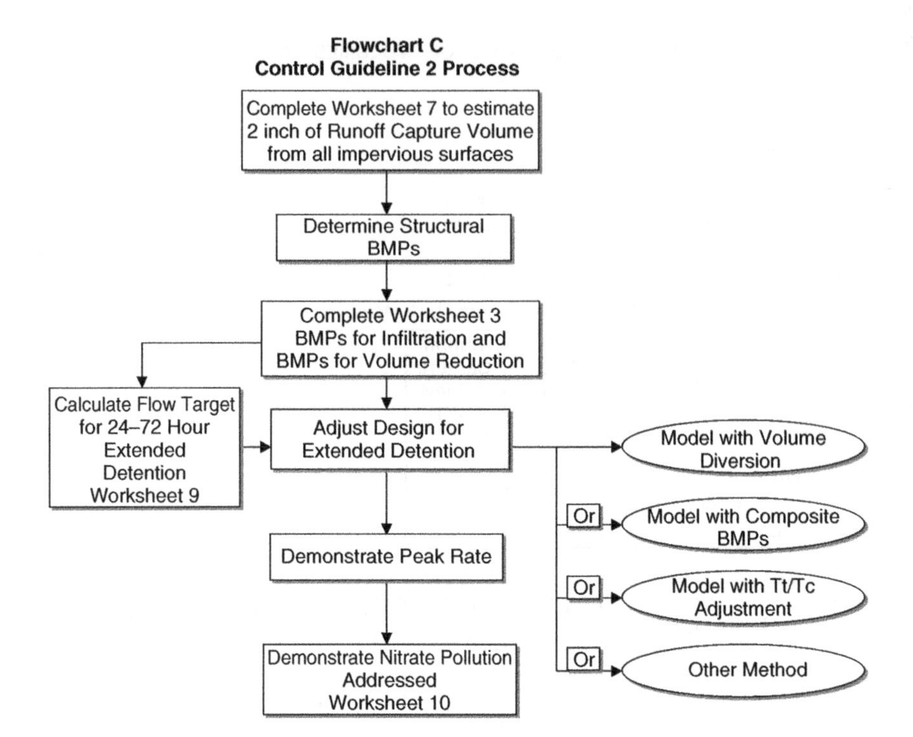
Flowchart C
Control Guideline 2 Process
Complete Worksheet 7 to estimate 2 inch of Runoff Capture Volume from all impervious surfaces
Determine Structural BMPs
Complete Worksheet 3 BMPs for Infiltration and BMPs for Volume Reduction
Calculate Flow Target for 24–72 Hour Extended Detention Worksheet 9
Adjust Design for Extended Detention
Model with Volume Diversion
Or
Model with Composite BMPs
Or
Model with Tt/Tc Adjustment
Or
Other Method
Demonstrate Peak Rate
Demonstrate Nitrate Pollution Addressed Worksheet 10

WORKSHEET 8 . STRUCTURAL BMP VOLUME CREDITS

PROJECT: ____________________

SUB-BASIN: ____________________

Required Control Volume (ft^3) - *from Worksheet 7 :* ____________________

Non-structural Volume Credit (ft^3) - *from Worksheet 3:* - ____________________

Structural Volume Reqmt (ft^3)
(*Required Control Volume minus Non-structural Credit*) ____________________

Proposed BMP*		Area (ft^2)	Storage Volume (ft^3)
6.4.1	Porous Pavement		
6.4.2	Infiltration Basin		
6.4.3	Infiltration Bed		
6.4.4	Infiltration Trench		
6.4.5	Rain Garden/Bioretention		
6.4.6	Dry Well / Seepage Pit		
6.4.7	Constructed Filter		
6.4.8	Vegetated Swale		
6.4.9	Vegetated Filter Strip		
6.4.10	Berm		
6.5.1	Vegetated Roof		
6.5.2	Capture and Re-use		
6.6.1	Constructed Wetlands		
6.6.2	Wet Pond / Retention Basin		
6.6.3	Dry Extended Detention Basin		
6.6.4	Water Quality Filters		
6.7.1	Riparian Buffer Restoration		
6.7.2	Landscape Restoration / Reforestation		
6.7.3	Soil Amendment		
6.8.1	Level Spreader		
6.8.2	Special Storage Areas		
Other			

Total Structural Volume (ft^3): ____________

Structural Volume Requirement (ft^3): ____________

DIFFERENCE ____________

WORKSHEET 9. CAPTURE / REUSE WATER BALANCE

PROJECT ____
SUB-BASIN: ____

Design Storage (gal):	10000

				Supply	Demand									
Month	Average Rainfall (in)	Capture/ Reuse Area (sf)	Runoff Coefficient of Capture/ Reuse Area	Average Monthly Runoff Volume (gal)	Monthly Irrigation Demand (gal)	Monthly Interior Water Demand (gal)	Monthly Total Water Demand (gal)	Deficit (gal)	Begin Month (gal)	Storage at Start of Month (gal)	End Month (gal)	Storage at End of Month (gal)	Overflow Volume (gal)
January	2	20000	0.9	22440	0	5000	5000	0	10000	10000	10000	10000	17440
February	2	20000	0.9	22440	0	5000	5000	0	10000	10000	10000	10000	17440
March	1	20000	0.9	11220	0	5000	5000	0	10000	10000	10000	10000	6220
April	1	20000	0.9	11220	10000	5000	15000	3780	10000	10000	6220	6220	0
May	0.5	20000	0.9	5610	10000	5000	15000	9390	6220	6220	−3170	0	0
June	0.5	20000	0.9	5610	10000	5000	15000	9390	−3170	0	−9390	0	0
July	0.1	20000	0.9	1122	10000	5000	15000	13878	−9390	0	−13878	0	0
August	0.5	20000	0.9	5610	10000	5000	15000	9390	−13878	0	−13878	0	0
September	1	20000	0.9	11220	10000	5000	15000	3780	−9390	0	−3780	0	0
October	1	20000	0.9	11220	0	5000	5000	0	−3780	0	6220	6220	2440
November	2	20000	0.9	22440	0	5000	5000	0	6220	6220	10000	10000	17440
December	2	20000	0.9	22440	0	5000	5000	0	10000	10000	10000	10000	17440

RUNOFF

Total Runoff Volume (gal)	152592
Overflow Volume (gal)	78420
Runoff Volume Captured (gal)	74172
% of Runoff Captured	49%

WATER SUPPLY

Total Demand (gal)	120000
Runoff Volume Captured (gal)	74172
% Supply Met	62%

Guidelines: Volume Credits for Detention Routing

(For projects required to demonstrate peak rate control / extended detention for 1-year through 100-year storm events)

The utilization of volume reduction BMPs as part of either CG-1 or CG-2 will obviously reduce the amount of storage required for peak rate mitigation because less runoff is ultimately discharged. As quantifying the peak rate mitigation benefits of volume control BMPs can be difficult and cumbersome with common stormwater models, applicants are strongly encouraged to use these or other approaches to give credit for volume control when demonstrating peak rate control:

1) Volume Diversion. Many computers models have components that allow a "diversion" or "abstraction". The total volume reduction provided by the applicable structural and non-structural BMPs can be diverted or abstracted from the modeled runoff before it is routed to the detention system(s). This approach is very conservative because it does not give any credut to the increased time of travel, ongoing infiltration, etc. associated with the BMPs. Incorporating the CG-2 capture volume in the model reduces the required detention storage by about 22% for this example.

2) Composite BMPs. For optimal stormwater management, this manual suggests widely distributed BMPs for volume, rate, and quality control. This approach, however, can be very cumbersome to evaluate in detail with common computer models. To facilitate modeling, similar types of BMPs can be combined within the model. For modeling purposes, the storage of the combined BMP is simply the sum of the BMP capacities that it represents. A stage-storage-discharge relationship can be developed for the combined BMP based on the configuration of the individual systems. The combined BMPs can then be routed normally and the results submitted.

3) Travel Time/ Time of Concentration Adjustment. The use of widely-distributed, volume-reducing BMPs can significantly increase the post-development runoff travel time and therefore decrease the peak rate of discharge. The Delaware Urban Runoff Management Model (DURMM) calculates the extended travel time through storage elements, even at flooded depths, to adjust peak flow rates (Lucas, 2001). The extended travel time is essentially the residence time of the storage elements, found by dividing the total storage by the 2-year peak flow rate. This increased travel timecan be added to the time of concentration of the area to account for the slowing effect of the volume-reducing BMPs. This can reduce the amount of detention storage required for peak rate control. The detention storage requirements are reduced by almost 40% by the CG-2 capture volume.

4) Other Methods. Other methods, such as adjusting runoff curve numbers based on the runoff volume left after BMP application, or reducing net precipitation based on the volume captured, can be applied as appropriate.

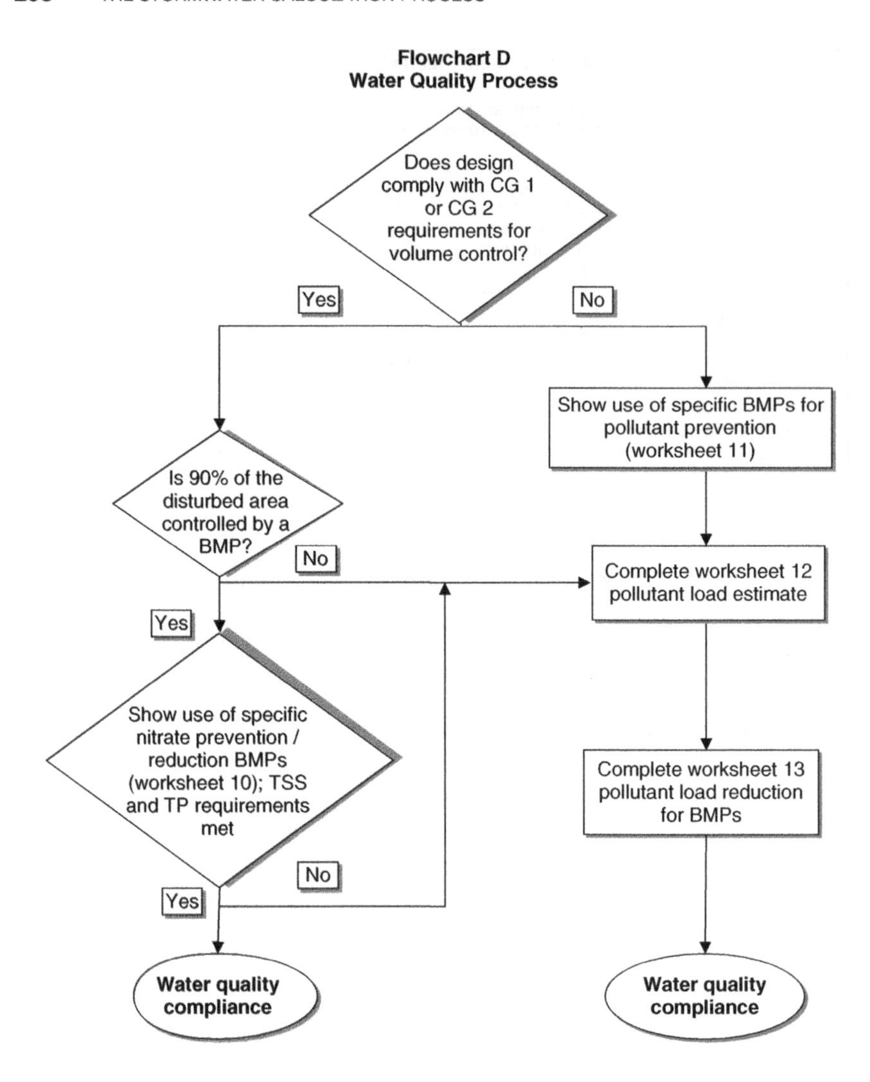
Flowchart D
Water Quality Process
Does design comply with CG 1 or CG 2 requirements for volume control?
Yes
No
Show use of specific BMPs for pollutant prevention (worksheet 11)
Is 90% of the disturbed area controlled by a BMP?
No
Complete worksheet 12 pollutant load estimate
Yes
Show use of specific nitrate prevention / reduction BMPs (worksheet 10); TSS and TP requirements met
Complete worksheet 13 pollutant load reduction for BMPs
No
Yes
Water quality compliance
Water quality compliance

WORKSHEET 10. WATER QUALITY COMPLIANCE FOR NITRATE

Does the site design incorporate the following BMPs to address nitrate pollution? A summary "yes" rating is achieved if at least 2 Primary BMPs for nitrate are provided across the site or 4 secondary BMPs for nitrate are provided across the site (or the

PRIMARY BMPs FOR NITRATE:

	YES	No
NS BMP 5.4.2 – Protect / Conserve / Enhance Riparian Buffers		
NS BMP 5.5.4 – Cluster Uses at Each Site		
NS BMP 5.6.1 – Minimize Total Disturbed Area		
NS BMP 5.6.3 – Re-Vegetate / Re-Forest Disturbed Areas (Native Species)		
NS BMP 5.9.1 – Street Sweeping / Vacuuming		
Structural BMP 6.7.1 – Riparian Buffer Restoration		
Structural BMP 6.7.2 – Landscape Restoration		

SECONDARY BMPs FOR NITRATE:

	YES	No
NS BMP 5.4.1 – Protect Sensitive / Special Value Features		
NS BMP 5.4.3 – Protect / Utilize Natural Drainage Features		
NS BMP 5.6.2 – Minimize Soil Compaction		
Structural BMP 6.4.5 – Rain Garden / Bioretention		
Structural BMP 6.4.8 – Vegetated Swale		
Structural BMP 6.4.9 – Vegetated Filter Strip		
Structural BMP 6.6.1 – Constructed Wetland		
Structural BMP 6.7.1 – Riparian Buffer Restoration		
Structural BMP 6.7.2 – Landscape Restoration		
Structural BMP 6.7.3 – Soils Amendment/Restoration		

WORKSHEET 11. BMPS FOR POLLUTION PREVENTION

Does the site design incorporate the following BMPs to address nitrate pollution? A summary "yes" rating is achieved if at least 2 BMPs are provided across the site. "Provided across the site" is taken to mean that the specifications for that BMP set fo

BMPs FOR POLLUTANT PREVENTION:

BMP	YES	No
NS BMP 5.4.1 - Protect Sensitive / Special Value Features		
NS BMP 5.4.2 - Protect / Conserve / Enhance Riparian Buffers		
NS BMP 5.4.3 - Protect / Utilize Natural Flow Pathways in Overall Stormwater Planning and Design		
NS BMP 5.5.1 - Cluster Uses at Each Site; Build on the Smallest Area Possible		
NS BMP 5.6.1 - Minimize Total Disturbed Area - Grading		
NS BMP 5.6.2 - Minimize Soil Compaction in Disturbed Areas		
NS BMP 5.6.3 - Re-Vegetate / Re-Forest Disturbed Areas (Native Species)		
NS BMP 5.7.1 - Reduce Street Imperviousness		
NS BMP 5.7.2 - Reduce Parking Imperviousness		
NS BMP 5.8.1 - Rooftop Disconnection		
NS BMP 5.8.2 - Disconnection from Storm Sewers		
NS BMP 5.9.1 - Street Sweeping		
Structural BMP 6.7.1 - Riparian Buffer Restoration		
Structural BMP 6.7.2- Landscape Restoration		
Structural BMP 6.7.3- Soils Amendment and Restoration		

WORKSHEET 12. WATER QUALITY ANALYSIS OF POLLUTANT LOADING FROM ALL DISTURBED AREAS

TOTAL SITE AREA (AC)	
TOTAL DISTURBED AREA (AC)	
DISTURBED AREA CONTROLLED BY BMPs (AC)	

TOTAL DISTURBED AREAS:

		POLLUTANT					POLLUTANT LOAD		
	LAND COVER CLASSIFICATION	TSS EMC (mg/l)	TP EMC (mg/l)	Nitrate-Nitrite EMC (mg/l as N)	COVER (Acres)	RUNOFF VOLUME (AF)	TSS** (LBS)	TP** (LBS)	NO^3 (LBS)
Pervious Surfaces	Forest	39	0.15	0.17					
	Meadow	47	0.19	0.3					
	Fertilized Planting Area	55	1.34	0.73					
	Native Planting Area	55	0.40	0.33					
	Lawn, Low-Input	180	0.40	0.44					
	Lawn, High-Input	180	2.22	1.46					
	Golf Course Fairway/Green	305	1.07	1.84					
	Grassed Athletic Field	200	1.07	1.01					
Impervious Surfaces	Rooftop	21	0.13	0.32					
	High Traffic Street / Highway	261	0.40	0.83					
	Medium Traffic Street	113	0.33	0.58					
	Low Traffic / Residential Street	86	0.36	0.47					
	Res. Driveway, Play Courts, etc.	60	0.46	0.47					
	High Traffic Parking Lot	120	0.39	0.60					
	Low Traffic Parking Lot	58	0.15	0.39					
						TOTAL LOAD			
						REQUIRED REDUCTION (%)	85%	85%	50%
						REQUIRED REDUCTION (LBS)			

* Pollutant Load = [EMC, mg/l] X [Volume, AF] X [2.7, Unit Conversion]

** TSS and TP calculations only required for projects not meeting CG1/CG2 or not controlling less than 90% of the disturbed area

APPENDIX B

CASE STUDIES

B.1 THE TRANSITION FROM RESEARCH TO PRACTICE

Over a period of 34 years, Cahill Associates (or Resource Management Associates) has participated in and conducted a number of studies and/or design projects, frequently in partnership with consultants in other fields (especially landscape architecture and green roof design) that offer a chronology of the evolution of the body of practice that we have come to call low-impact development and sustainable stormwater management. These studies and designs illustrate how the initial concerns of water quality impacts resulting from runoff-transported pollutants (especially sediment) have gradually widened to include other aspects of the land development process: social, environmental, and economic. Professional practitioners now include engineers, planners, architects, landscape architects, economists, developers, financiers, and government regulators, with very different perspectives on the subject of LID.

In the late 1960s and early 1970s, the neglected water quality issue was sediment pollution, unregulated in all federal, state, and local regulatory processes and permits. The national agency that had addressed this issue for decades in agriculture [the U.S. Soil Conservation Service (the SCS), now the NRCS] urged other regulators to adopt requirements for control of soil erosion from land development. The solution proposed was based on the traditional farm pond of the 1930s, which had helped to stem the soil loss from much of the nation's cultivated lands. The sediment running from every land development site was considered to be just a variation of the problem, and the farm pond was changed to include

Low Impact Development and Sustainable Stormwater Management, First Edition. Thomas H. Cahill.

an outlet that detained runoff and mitigated the peak flow for a sufficient period to avoid downstream bank erosion. A full set of design guidelines evolved and were promulgated, usually at the local or county level rather than being incorporated in state or federal water quality regulations, in order to influence the land development process in concert with zoning and development regulations. The suburban development explosion that was taking place throughout the nation was suddenly faced with a technical requirement to control runoff, a practice that spread rapidly to every community.

Within an incredibly short period of a few years, every new land development project included a detention basin, following sizing and performance criteria recommended by the SCS and tailored in each community to fit local concerns and practices. It would be accurate to say that the detention basin became the national standard for stormwater management in the land development process, and has continued to serve as such for the past 35 years. When the technical community of water quality experts began to make the connection between sediment pollution as the transport mechanism for phosphorus and other pollutants, and the root cause of eutrophication of almost every freshwater body in developed regions, the need to control (or prevent) this pollutant load from the landscape reaching these surface waters became critical.

Even as hundreds of thousands of detention basins were constructed during the 1970s and 1980s, it became apparent that this measure, in itself, was insufficient to offset the full impact of the land development process on water resources. Much more needed to be done in the actual site construction process, by reducing the amount of land disturbed, reducing the amount of earthwork, configuring dwelling units in closer patterns, and not simply counting on the detention basin to capture runoff and prevent sediment pollution. During the 1980s, various investigators (including the present author) advocated that the primary technical solution to preventing increased runoff from developed landscapes was to focus on volume control rather than mitigating the rate of runoff as accomplished by detention designs. Despite the initial rejection of this concept, various technologies using innovative methods and materials were constructed. These included porous pavements (first tested in the 1970s and finding broad application in the 1980s), vegetated roofs (derived from the German experience of the late 1970s and 1980s and brought to the United States in the 1990s), and vegetative systems that were designed in the 1980s as "vegetated infiltration beds" but caught on as "rain gardens" in the 1990s. Other technologies, especially capture–reuse systems, were used in other countries but met with resistance in the United States, based largely on existing water supply infrastructure (and ownership), fears of supply contamination, and regulatory obstacles.

The case studies included here are divided into three categories: manuals, basin plans or models, and constructed LID projects. While the basic issue of runoff-transported pollution is the central concern throughout, the popular acronym has evolved and changed over the 35-year period, from non-point Source (NPS) control to best management practices (BMPs) to low-impact development (LID). The more recent term *living building* goes beyond the site design issues and has more to do with the building structure itself and the water-energy aspects of the structure, although LID solutions are an important part of the water issues. The three manuals discussed initially reflect the changing theme of NPS, BMPs, and LID, as the designs moved into the general construction and land development market in an organized fashion. The planning studies also reflect this transition from theory to practice. The design case studies are similar in form but show the application in different environments and physiographic regions, and are drawn from several hundred built projects.

B.2 MANUALS

A manual is intended as a guidance document for practioners of a given profession to illustrate various methods of both design and construction. Since this particular subject began from a basis of water quality concerns, and all of the initial legislative and regulatory criteria were expressed in those terms, it is appropriate to illustrate how such manuals have evolved over the past 20 years. The documents included are not intended to summarize all of the various manuals prepared over that period, only to show how the same basic problem has produced solutions that have evolved from solving a pollution issue to building a more sustainable site.

EPA Non-Point Source Pollution Control Manual (1992)

In the Coastal Zone Act Reauthorization Amendments of 1991, Congress directed both the Environmental Protection Agency (EPA) and the National Oceanic and Atmospheric Administration (NOAA) to develop new programs to control non-point sources (NPSs) of pollution (Section 6217). Congress stipulated that EPA and NOAA must give to all 30 states with coastal waters minimum guidance to establish new pollution control programs. Cahill Associates served as part of the nationwide team of consultants assisting in the development of this program, specifically for urban growth–related pollution sources within coastal drainage. The end result was a non-point source pollution management manual entitled *Guidance Specifying Management Measures for Sources of Nonpoint Pollution in Coastal Waters* (Figure B-1), a massive document that represented a compendium of both agricultural and urban NPS.

Cahill Associates had previously performed several studies of water quality impacts in estuarine systems along the New Jersey coast and had developed nonstructural measures for prevention of pollution rather than mitigation with structural measures. A technique developed for the New Jersey coastal zone management program by Cahill Associates, called minimum disturbance/minimum maintenance land development, was incorporated in the manual. These site development management measures required that developers plan, design, and develop every site to:

1. Protect areas that provide important water quality benefits and/or are particularly susceptible to erosion and sediment loss.

Guidance Specifying Management Measures For Sources Of Nonpoint Pollution In Coastal Waters

Issued Under the Authority of Section 6217(g) of the Coastal Zone Act Reauthorization Amendments of 1990

United States Environmental Protection Agency
Office of Water
Washington, DC

Figure B-1 First national effort to develop an NPS manual (EPA/NOAA).

2. Limit increases of impervious areas, except where necessary.
3. Limit land disturbance activities such as clearing and grading, and cut and fill to reduce erosion and sediment loss.
4. Limit disturbance of natural drainage features and vegetation.

The end result of many months of discussion by several dozen experts was a massive compendium of papers on a broad range of related (and unrelated) subjects, some of which were in conflict with each other. By and large, the detention systems received the most treatment, and the volume-reduction technologies were given only passing consideration. As an initial effort to pull together the collective experience of reducing NPS pollution, it did serve to demonstrate the state of understanding after almost two decades of effort, and led to much better efforts to prepare a guidance document by a number of states over the following decade. This document was published by U.S. EPA, Office of Water, Washington, DC 20460, in January 1992.

During the 1990s, a number of states took the initiative and prepared BMP manuals listing and explaining in some detail the various types of erosion control, sediment capture, and rate mitigation measures, again trying to design a better detention basin, now including vegetative elements and, where possible, infiltration elements.

Pennsylvania BMP Manual (2005)

The commonwealth of Pennsylvania followed the trend of other states, preparing a BMP manual during the late 1990s directed at the local units of government (2,560 throughout the state), not as regulations (because the issue of land use was still considered a local issue and politically sensitive) but as "guidance." Like other state manuals, this largely failed to change the development process, and the building of detention basins continued unabated as the appropriate stormwater management system, augmented by a number of erosion control technologies to contain the soil loss within the development site.

From 2003 through 2005, Cahill Associates helped the Pennsylvania Department of Environmental Protection (PADEP) develop the *Pennsylvania Stormwater Best Management Practices Manual* (Figure B-2). Although a number of states and other jurisdictions had prepared BMP manuals, this was the first state manual to aggressively advocate volume-reduction technologies, although not without great discussion. An advisory committee was structured to assist Cahill Associates and the PADEP in this brainstorming process, and although the final design criteria were by no means embraced uniformly by all parties, the BMP manual did represent the closest effort to structuring the concepts of low-impact development.

The most important recommendation to come out of this work was the design criterion for volume control, expressed as follows:

> All new development shall prevent any net increase in the volume of runoff generated from a development site during the 2-year, 24-hour frequency rainfall.

Multiple criteria were discussed and considered, and nonstructural measures were strongly recommended, but again the subject of local government control of the land development process influenced the final document. This manual has served as the basis for inclusion in many local land development regulations, but because of the rural nature of much of the state, has yet to find full adoption in all of the 2,560 municipalities across the state.

The major urban centers, including Philadelphia and Pittsburgh, were struggling with a multiplicity of land redevelopment issues, only one of which was stormwater. In communities with combined sewers the impact was felt in the uncontrolled overflow from combined sewer overflows (CSOs) to surface waters, a much more direct water quality impact and health threat, especially to communities situated downstream. For these urban environments, independent

Table of Contents

Pennsylvania Stormwater Best Management Practices Manual

2ND Draft - JANUARY 2005

Under Contract No. 2003-WWEC-1

Prepared By:

Cahill Associates Inc.
104 S. High Street
West Chester, PA 19382
(610) 696-4150

With assistance from:

Amy Green Associates
Flemington, NJ

GeoSyntec Consultants
Boxborough, MA

Low Impact Development Center, Inc.
Rockville, MD

Pennsylvania Environmental Council
Philadelphia, Harrisburg, Pittsburgh, PA

RoofScapes, Inc.
Philadelphia, PA

Figure B-2 The second Pennsylvania BMP manual was structured on LID Principles.

stormwater management approaches were developed during the 1990s and into the 2000s, especially in Philadelphia. The end result of this effort is very close to the LID program advocated statewide, although the design criteria used were intended to control sufficient runoff volume to prevent CSO discharges and are less strict numerically.

B.3 LID MANUAL FOR MICHIGAN (2008)

In 2007–2008, Cahill Associates assisted the Southeast Michigan Council of Governments (SEMCOG) in the preparation of the LID manual for the state of Michigan. Focused initially on the six counties in the southeastern portion of the state, which is the most highly urbanized portion of the state and includes the metropolitan region of Detroit. The manual was supported by a state grant which required that the guidance be representative of the entire state, and set the bar for both new development and redevelopment, an important issue in the older urban communities.

This manual focused exclusively on volume-reduction designs and was tailored to the specific hydrology of Michigan, a glacially sculpted landform that produces a regional hydrology with low base flow and significant natural soil infiltration in many regions. The best example of this approach is contained in Chapter 2 of this manual, included here as the most comprehensive LID manual developed as of 2008. The writing style (largely by SEMCOG staff) is produced in a form easily understood by the nontechnical community and is well illustrated in color. It is also available in electronic form. It is far superior to earlier manuals.

Chapter 2

Stormwater Management in Michigan: Why LID?

Clean water resources are essential to the economic vitality of Michigan. Proper stormwater management is an essential component of water quality protection. Low impact development is a cornerstone of stormwater management and thus is the pathway to protecting water resources and enabling economic growth.

This chapter discusses:

- The importance of the water cycle
- The impacts of stormwater runoff
- An overview of what LID is and how it works
- Benefits of implementing LID
- Cost effectiveness and LID
- Relationship of LID to other programs
- Getting started with LID.

The importance of the water cycle

A key component of protecting water resources is keeping the water cycle in balance. The movement of rainfall from the atmosphere to the land and then back to the atmosphere—the water (hydrologic) cycle—is a naturally continuous process essential to human and virtually all other forms of life (Figure 2.1). This balanced water cycle of precipitation, evapotranspiration, infiltration, groundwater recharge, and stream base flow sustains Michigan's vast but fragile water resources.

In a natural woodland or meadow in Michigan, most of the annual rainfall soaks into (infiltrates) the soil mantle. Over half of the annual rainfall returns to the atmosphere through evapotranspiration. Surface vegetation, especially trees, transpire water to the atmosphere with seasonal variations.

Water that continues to percolate downward through the soil reaches the water table and moves slowly downgradient under the influence of gravity, ultimately providing baseflow for streams and rivers, lakes, and wetlands. On an annual basis, under natural conditions, only a small portion of annual rainfall results in immediate stormwater runoff (Figure 2.2). Although the total amount of rainfall varies in different regions of the state (see Chapter 3), the basic relationships of the water cycle are relatively constant.

Figure 2.1 Water Cycle. *Source*: Stream Corridor Restoration: Principles. Processes, and Practices, 10/98, Federal Interagency Stream Restoration Working Group (FISRWG).

Conventional land development changes the land surface and impacts the water cycle (Figure 2.3). Altering one component of the water cycle invariably causes changes in other elements of the cycle. Impervious surfaces, such as roads, buildings, and parking areas, prevent rainfall from soaking into the soil and significantly increase the amount of rainfall that runs off. Additionally, research shows that soil compaction resulting from land development produces far more runoff than the presettlement soil conditions. As natural vegetation systems are removed, the amount of evapotranspiration decreases. As impervious areas increase, runoff increases, and the amount of groundwater recharge decreases.

These changes in the water cycle have a dramatic effect on our water resources. As impervious and disturbed or compacted pervious surfaces increase and runoff volumes increase, stream channels erode, substrate in the river

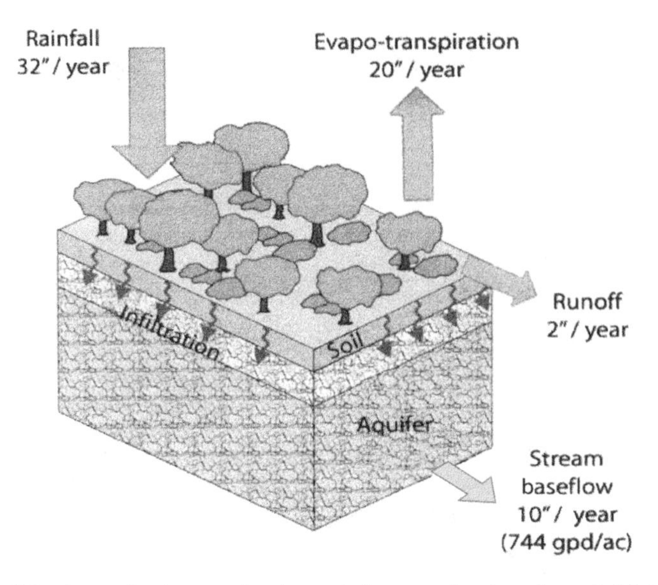

Figure 2.2 Approximate annual water cycle for an undeveloped acre in Michigan.

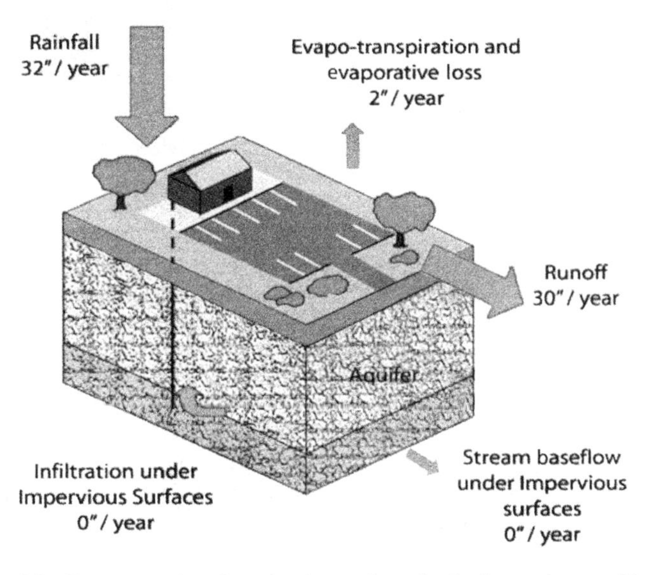

Figure 2.3 Representative altered water cycle under the impervious parking lot.

bottom is impacted, habitat is lost or reduced, and populations of fish and other aquatic species decline. Reduced infiltration and groundwater recharge results in lowered water tables and reduced stream baseflow, generally worsening low flow conditions in streams during dry periods.

The impacts of stormwater runoff

Stormwater runoff is rainfall or snowmelt that runs off the land and is released into rivers and lakes. Problems related to stormwater runoff are most evident in areas where urbanization has occurred. As mentioned above, the change in the water cycle has a dramatic effect on our water resources. This impact is based on both the quantity and quality of stormwater runoff reaching our rivers and lakes.

The impacts of stormwater runoff are well documented in Michigan and throughout the country. They include:

- **Increased flooding and property damage.** Increased impervious surfaces decrease the amount of rainwater that can naturally infiltrate into the soil and increase the volume and rate of stormwater runoff. These changes lead to more frequent and severe flooding and potential damage to public and private property.
- **Degradation of the stream channel.** One result of runoff can be more water moving at higher velocities through stream channels. This condition is called "flashy flows" and happens at increased frequency as an area is developed. As a result, both the streambank and stream bed are eroded more frequently. This can result in widening and deepening the channel, as well as a decline in stream substrate quality, and degradation of habitat.
- **Less groundwater recharge and dry weather flow.** As impervious surfaces increase, the infiltration of stormwater to replenish groundwater decreases. Groundwater is important because many people rely on groundwater for their drinking water supply. In addition, the groundwater "feeds" rivers and lakes especially during the dry season to ensure a steady flow.

Streambank erosion and degraded habitat. *Source*: Wayne County Department of Environment.

When the groundwater recharge decreases, the amount of dry weather flow decreases, negatively impacting aquatic life and recreational opportunities.

- **Impaired water quality.** Impervious surfaces accumulate pollutants that are absorbed by stormwater runoff and carried to lakes and streams. Examples of these pollutants include:
  - Hydrocarbons and trace metals from vehicles.
  - Suspended solids from erosive stream banks and construction sites.
  - Chlorides from road salt.
  - Nutrients from fertilizer and grass clippings and leaves left on streets and sidewalks, and
  - Bacteria from pet waste, goose droppings, and other wildlife.
- **Increased water temperature.** Impervious surfaces are warmed by the sun. Runoff from these warmed surfaces increase the temperature of water entering our rivers and lakes. This can adversely impact aquatic life that requires cold water conditions (e.g., trout).
- **Loss of habitat.** The decline in habitat due primarily to the erosive flows and the increased water temperature will negatively impact the diversity and amount of fish and aquatic insects.
- **Decreased recreational opportunities.** Stormwater runoff can negatively impact water resources in many different ways (e.g., decreased water quality, increased temperature, and decreased habitat). The result is diminished recreational and economic opportunities for communities throughout the state.

Stormwater solutions — low impact development

What is LID?

From a stormwater management perspective, low impact development (LID) is the application of techniques that emulate the natural water cycle described in the previous section LID uses a basic principle modeled after nature: manage rainfall by using design techniques that infiltrate, filter, store, evaporate, and detain runoff close to its source.

Techniques are based on the premise that stormwater is a resource, not a waste to be quickly transported and disposed. Instead of conveying and managing/treating stormwater in large, costly, end-of-pipe facilities located often at the bottom of drainage areas, LID addresses stormwater through small, cost-effective landscape features often located at the lot level.

Almost all components of the urban environment have the potential to serve as elements of an integrated stormwater management system. This includes open space, as well as rooftops, streetscapes, parking lots, sidewalks, and medians. LID is a versatile approach that can be applied equally well to new

Native plantings at East Grand Rapids, MI Community Center.

development, urban redevelopment, and in limited space applications such as along transportation corridors.

How does LID work?

LID strives to replicate virtually all components of the natural water cycle by:

- Minimizing total runoff volume
- Controlling peak rate of runoff
- Maximizing infiltration and groundwater recharge
- Maintaining stream baseflow
- Maximizing evapotranspiration
- Protecting water quality.

Stormwater management historically focused on managing the flood effects from larger storms. Exclusive reliance on peak rate control prevents flooding, but doesn't protect streams and water quality. Thorough stormwater management should target infrequent large storms, as well as the much more frequent, smaller storms.

With the change in land surface generated by land development, not only does the peak rate of runoff increase, but the *total volume* of runoff also often dramatically increases. LID focuses on both peak rates and total volumes of runoff. LID application techniques are designed to hold constant peak rates of runoff for larger storms and prevent runoff volume increases for the much more frequent, smaller storms. Thus, the natural flow pattern is kept in better balance, avoiding many of the adverse impacts associated with stormwater runoff.

Table 2.1 Pollutant Removal Table (in percentages)

Pollutant	Infiltration Practices	Stormwater Wetlands	Stormwater Ponds Wet	Filtering Practices	Water Quality Swales	Stormwater Dry Ponds
Total Phosphorus	70	49	51	59	34	19
Soluble Phosphorus	85	35	66	3	38	−6
Total Nitrogen	51	30	33	38	84	25
Nitrate	82	67	43	−14	31	4
Copper	N/A	40	57	49	51	26
Zinc	99	44	66	88	71	26
TSS	95	76	80	86	81	47

Source: "National Pollutant Removal Performance Database for Stormwater Treatment Practices". Center for Watershed Protection, June 2000.

LID focuses on the following stormwater outcomes, described in more detail in Chapter 9:

- Preventing flooding
- Protecting the stream channel
- Improving and protecting water quality
- Recharging groundwater.

Chapter 9 describes recommended criteria that communities and/or developers may use at the site level to implement LID designs. This may also be used at the community level to develop standards to ensure that development meets the outcomes listed above.

Infiltration practices often associated with LID provide enhanced water quality benefit compared to many other BMPs. Percent of pollutant removal for various LID practices is shown in the table below.

Principles of LID

Successful application of LID is maximized when it is viewed in the context of the larger design process. This process is reflected in a set of principles used to guide development of this manual.

- Plan first
- Prevent, then mitigate
- Minimize disturbance
- Manage stormwater as a resource — not a waste
- Mimic the natural water cycle

- Disconnect. Decentralize. Distribute
- Integrate natural systems
- Maximize the multiple benefits of LID
- Use LID everywhere
- Make maintenance a priority.

Plan first. To minimize stormwater impacts and optimize the benefits of LID, stormwater management and LID should be integrated into the community planning and zoning process.

Prevent. Then mitigate. A primary goal of LID is preventing stormwater runoff by incorporating nonstructural practices into the site development process. This can include preserving natural features, clustering development, and minimizing impervious surfaces. Once prevention as a design strategy is maximized, then the site design — using structural BMPs — can be prepared.

Minimize disturbance. Limiting the disturbance of a site reduces the amount of stormwater runoff control needed to maintain the natural hydrology.

Manage stormwater as a resource — not a waste. Approaching LID as part of a larger design process enables us to move away from the conventional concept of runoff as a disposal problem (and disposed of as rapidly as possible) to understanding that stormwater is a resource for groundwater recharge, stream base flow, lake and wetland health, water supply, and recreation.

Mimic the natural water cycle. Stormwater management using LID includes mimicking the water cycle through careful control of peak rates as well as the volume of runoff and groundwater recharge, while protecting water quality. LID reflects an appreciation for management of both the largest storms, as well as the much more frequent, smaller storms.

Disconnect. Decentralize. Distribute. An important element of LID is directing runoff to BMPs as close to the generation point as possible in patterns that are decentralized and broadly distributed across the site.

Integrate natural systems. LID includes careful inventorying and protecting of a site's natural resources that can be integrated into the stormwater management design. The result is a natural or "green infrastructure" that not only provides water quality benefits, but greatly improves appearance by minimizing infrastructure.

Maximize the multiple benefits of LID. LID provides numerous stormwater management benefits, but also contributes to other environmental, social, and economic benefits. In considering the extent of the application of LID, communities need to consider these other benefits.

Use LID everywhere. LID can work on redevelopment, as well as new development sites. In fact, LID can be used on sites that might not traditionally consider LID techniques, such as in combined sewer systems, along transportation corridors, and on brownfield sites. Broad application of LID techniques improves the likelihood that the desired outcome of water resource protection and restoration will be achieved.

Make maintenance a priority. The best LID designs lose value without commitment to maintenance. An important component of selecting a LID technique is understanding the maintenance needs and institutionalizing a maintenance program. Selection of optimal LID BMPs should be coordinated with both the nature of the proposed land use/building program and the owners/operators of the proposed use for implementation of future maintenance activities.

Benefits of implementing LID

Implementing LID offers numerous benefits to communities, developers, and the public that extend well beyond water quality protection. Here are some examples:

Communities, agencies, and the public

- Reduces municipal infrastructure and utility maintenance costs (e.g., streets, curbs, gutters, storm sewers).
- Increases energy and cost savings for heating, cooling, and irrigation.
- Reduces flooding and streambank erosion.
- Replenishes groundwater drinking supply.
- Assists in meeting regulatory obligations.
- Serves multiple purposes (e.g., traffic calming, greenways).
- Brings neighborhoods together in maintaining LID.
- Increases recreational opportunities.
- Provides environmental education opportunities.
- Improves quality of life for residents.
- Protects community character/aesthetics.
- Protects and enhances sensitive habitat.
- Restores/protects fisheries and other aquatic life.
- Reduces salt usage and snow removal on paved surfaces.

Developers

- Reduces land clearing and grading costs.
- Potentially reduces infrastructure costs (e.g., streets, curb, gutters).
- Reduces stormwater management construction costs.

- Increases marketability leading to faster sales.
- Potentially increases lot yields/amount of developable land.
- Assists in meeting LEED (Leadership in Energy and Environmental Design) Certification requirements.
- Appealing development consistent with the public's desire for environmental responsibility.

Recreation in Glen Haven, MI

Michigan inland lakeshore on Horseshoe Lake, Northfield Township, MI

Traverse City, MI, Marina

Environmental

- Protects/restores the water quality of rivers and lakes.
- Protects stream channels.
- Reduces energy consumption.
- Improves air quality.
- Preserves ecological and biological systems.
- Reduces impacts to terrestrial and aquatic plants and animals.
- Preserves trees and natural vegetation.
- Maintains consistent dry weather flow (baseflow) through groundwater recharge.
- Enhances carbon sequestration through preservation and planting of vegetation.

Cost effectiveness and LID

A variety of sources are now available documenting the cost effectiveness — even cost reductions — which can be achieved through the application of LID practices. The U.S. Environmental Protection Agency (EPA) released *Reducing Stormwater Costs Through Low Impact Development (LID) Strategies and Practices*, reporting on cost comparisons for 17 different case studies across the country. EPA results demonstrate the positive cost advantages of LID practices, when compared with traditional development patterns using conventional stormwater management techniques.

Based on this recent work, EPA concludes that, in the majority of cases, significant cost savings resulted from reduced site grading and preparation, less stormwater infrastructure, reduced site paving, and modified landscaping. Total capital cost savings ranged from 15 to 80 percent when using LID methods. Furthermore, these results are likely to conservatively undercount LID benefits. In all cases, there were benefits that this EPA study did not monetize or factor into each project's bottom line. These benefits include:

- Improved aesthetics,
- Expanded recreational opportunities,
- Increased property values due to the desirability of the lots and their proximity to open space,
- Increased total number of units developed,
- Increased marketing potential, and
- Faster sales.

Using LID to Meet Regulatory Requirements

LID practices can be used to meet a variety of state and federal permit programs. These range from the National Pollutant Discharge Elimination System (NPDES) Phase I and Phase II stormwater requirements, to combined sewer overflow (CSO) and sanitary sewer overflow (SSO) requirements. For example, many Michigan municipalities are plagued with CSO problems as well as SSOs caused by excessive inflow of stormwater and groundwater into the sanitary sewer system. Communities can integrate LID practices, such as a residential rain barrel program and downspout disconnection to their overflow control programs to help reduce stormwater inflow into the system, thereby reducing overflows.

Additionally, cost estimates do not include any sort of monetizing of the environmental impacts which are avoided through LID, as well as reductions in long-term operation and maintenance costs, and/or reductions in the life cycle costs of replacing or rehabilitating infrastructure.

Confirming EPA results, a recent report by the Conservation Research Institute for the Illinois Conservation Foundation. *Changing Cost Perceptions: An Analysis of Conservation Development*, 2005, undertook three different types of analyses on this cost issue—a literature review, an analysis of built-site case studies, and a cost analysis of hypothetical conventional versus conservation design templates. In terms of literature review, this study concludes:

- Public infrastructure costs are lower when a development is built within the context of smart growth patterns that conserve land.

Table 2.2 Summary of Cost Comparisons Between Conventional and LID Approaches

Project	Conventional Development Cost	LID Cost	Cost Difference	Percent Difference
2nd Avenue SEA Street	$868,803	$651,548	$217,255	25%
Aubum Hills	$2,360,385	$1,598,989	$761,396	32%
Bellingham City Hall	$27,600	$5,600	$22,000	80%
Bellingham Bloedel Donovan Park	$52,800	$12,800	$40,000	76%
Gap Creek	$4,620,600	$3,942,100	$678,500	15%
Garden Valley	$324,400	$260,700	$63,700	20%
Laurel Springs	$1,654,021	$1,149,552	$504,469	30%
Mill Creek[a]	$12,510	$9,099	$3,411	27%
Pairie Glen	$1,004,848	$599,536	$405,312	40%
Somerset	$2,456,843	$1,671,461	$785,382	32%
Tellabs Corporate Campus	$3,162,160	$2,700,650	$461,510	15%

Source: *Low Impact Development (LID) Strategies and Practices*, USEPA, 2007

[a]Mill Creek costs are reported on a per-lot basis.

- At the site level, significant cost savings can be achieved from clustering, including costs for clearing and grading, stormwater and transportation infrastructure, and utilities.
- Installation costs can be between $4,400 and $8,850 cheaper per acre for natural landscaping than for turf grass approaches.
- Maintenance cost savings range between $3,950 and $4,583 per acre, per year over 10 years for native landscaping approaches over turf grass approaches.
- While conventional paving materials are less expensive than conservation alternatives, porous materials can help total development costs go down, sometimes as much as 30 percent by reducing conveyance and detention needs.
- Swale conveyance is cheaper than pipe systems.
- Costs of retention or detention cannot be examined in isolation, but must instead be analyzed in combination with conveyance costs, at which point conservation methods generally have a cost advantage.
- Green roofs are currently more expensive to install than standard roofs, yet costs are highly variable and decreasing. Green roofs also have significant cost advantages when looking at life cycle costs (e.g., building, heating, and cooling costs).

Principles of Smart Growth

- Create a range of housing opportunities and choices.
- Create walkable neighborhoods.
- Encourage community and stakeholder collaboration.
- Foster distinctive, attractive communities with a strong sense of place.
- Make development decisions predictable, fair, and cost effective.
- Mix land uses.
- Preserve open space, farmland, natural beauty, and critical environmental areas.
- Provide a variety of transportation choices.
- Strengthen and direct development towards existing communities.
- Take advantage of compact building design.

Source: Smart Growth Network

Relationship of LID to other programs

LID is compatible with the principles of smart growth and the requirements of the U.S. Green Building Council's LEED program because LID offers prevention and mitigation benefits that make land development much more sustainable.

LID and Smart Growth

LID is often seen as a site specific stormwater management practice, while smart growth is often a broader vision held at a community, county, or regional level. However, as noted in Chapter 4, an important first step in LID is incorporating LID at the community level.

There are direct connections between LID and smart growth. For example, principles relating to compact building design and preserving natural features directly relate to nonstructural LID BMPs listed in Chapter 6. Upon further evaluation, LID is also consistent with the larger concepts of stakeholder collaboration; fostering communities with a strong sense of place; and implementing fair, predictable, and cost effective development decisions.

LID and LEED

The Leadership in Energy and Environmental Design (LEED) certification encourages and accelerates global adoption of sustainable green building and development practices by creating and implementing widely understood and accepted tools and performance criteria. LEED has developed rating systems for a myriad of development scenarios, including new construction, existing buildings, commercial interiors, core and shell, schools, retail, healthcare, homes, and neighborhood development.

As with Smart Growth, there are significant connections between LID and LEED certification. In fact, LID practices are integrated into each of the LEED rating systems.

Fairmount Square LEED Certification

Fairmount Square is a 4-acre infill site that uses rainwater capture, porous pavement, and rain gardens to manage its stormwater. The project is also seeking various LEED credits for new construction.

The building was designed with a focus of structural longevity and durability, energy efficiency, and a high quality indoor environment. Key site features include: better insulated concrete framing and roofing material and the use of low off-gassing interior materials such as carpet, paints, caulks, and adhesives. The project also takes advantage of existing infrastructure by being close to transit lines and other community features within walking distance to the site.

Fairmount Square, Grand Rapids, MI. Source: Fishbeck, Thompson, Carr & Huber, Inc.

The United States Green Building Council (USGBC), the Congress for New Urbanism and the National Resources Defense Council are currently working on a new rating system called LEED for Neighborhood Development (LEED-ND). The strongest connection between the LEED system and LID will be through LEED-ND certification. LEED-ND is part of the natural evolution of the green building movement, expanding sustainability standards to the scale of the neighborhood. While current green building standards focus on buildings in isolation, LEED-ND will bring emphasis to the elements that determine a development's relationship with its neighborhood, region, and landscape. LEED-ND sets standards in four categories that pinpoint essential neighborhood characteristics:

- Complete, compact, and connected neighborhoods
- Location efficiency
- Resource efficiency
- Environmental preservation.

Currently, the LEED-ND system is being piloted by the USGBC. The post-pilot version of the rating system, which will be available to the public, is expected to launch in 2009 (See LEED-ND criteria pullout).

Getting started with LID

LID can be implemented by many different groups, including communities, counties, developers, agencies, or individuals. Implementing LID can take many forms. For some, implementation might be encouraged on a voluntary basis during the site plan review process. For others, LID might become an expected application at each site and be institutionalized in an ordinance or through multiple ordinances.

A key first step is for different institutions within a local government to discuss the pros and cons of various approaches to LID. These stakeholders might include mayors/supervisors, councils/trustees, planning commissions,

City of Wixom, MI Habitat Park. Source: Hubbell, Roth & Clark. Inc.

public works department, etc. The outcome of these discussions will be action steps toward instituting LID at the desired scale on a community basis.

LEED-ND Criteria

Smart Location and Linkage (SLL)

SLL Prerequisite 3: Imperiled species and ecological communities
SLL Prerequisite 4: Wetland and water body conservation
SLL Prerequisite 6: Floodplain avoidance
SLL Credit 8: Steep slope protection
SLL Credit 9: Site design for habitat or wetland conservation
SLL Credit 10: Restoration of habitat or wetlands
SLL Credit 11: Conservation management of habitat of wetlands

Neighborhood pattern and design (NPD)

NPD Prerequisite 1: Open community
NPD Prerequisite 2: Compact development
NPD Credit 1: Compact development

Green construction and technology (GCT)

GCT Prerequisite 1: Construction activity pollution prevention
GCT Credit 3: Reduced water use
GCT Credit 6: Minimize site disturbance through site design
GCT Credit 7: Minimize site disturbance during construction
GCT Credit 9: Stormwater management
GCT Credit 10: Heat island reduction

References

Center for Watershed Protection. www.cwp.org

Conservation Research Institute. *Changing Cost Perceptions: An Analysis of Conservation Development.* 2005. www.nipc.org/environment/sustainable/conservation design/cost_analysis/

Smart Growth Network. www.smartgrowth.org

U.S. Environmental Protection Agency. *Reducing Stormwater Costs Through Low Impact Development (LID) Strategies and Practices.* December 2007. www.epa.gov/owow/nps/lid/costs07/

U.S. Green Building Council. LEED Rating System. www.usgbc.org

B.4 MODELS AND WATERSHED STUDIES

By the 1960s it was apparent that the nation's water resources were experiencing a significant reduction in quality. The root causes of that diminution in both rivers and lakes was attributed to nutrients produced by the thousands of wastewater treatment facilities then in place, serving most major cities and towns, and the inadequacy of the treatment processes used. After lengthy debate it was concluded that a major investment was needed to upgrade the efficiency of wastewater treatment process, a process that could be accomplished only with significant help from the federal government for municipalities and stringent requirements for industries. This public investment took the form of PL 92–500, passed in 1992, and initiated a massive effort that carried through the next two decades.

The issue of pollutants conveyed to surface waters from the land surface, especially agriculture, was recognized, but the resulting problems were first observed in lakes, where the influx of nutrients, especially phosphorus, was changing the trophic state of every impoundment. Pioneering research in the Tennessee Valley Authority system of lakes, built to produce hydroelectric energy for a portion of the country not well served by the national power grid, demonstrated clearly that this change in water quality could be attributed to the sediment and nutrients carried with every significant rainfall from the surrounding watershed. Studies in Lake Tahoe, situated on the border of California and Nevada, resulted in the recognition that any solution depended on better managing the land use of the surrounding drainage basin, and a planning process was initiated that continues today.

This research was followed by the passage of PL 92–500, and included in that act was specific wording on the issue of pollutant generation from land runoff and the impact on lakes and human-made impoundments. The center of attention was now Lake Erie, the shallowest of the Great Lakes, which had shown signs of early pollution and changes in trophic state, again attributed to the influx of phosphorus from major cities that ring the lake, such as Detroit, Cleveland, Toledo, Buffalo, and Sandusky. It was not clear from the available data exactly how much phosphorus was coming from these treatment facilities and how much was coming from the surrounding 23,000-m^2 drainage area, much of which was in active cultivation.

Following these studies, national attention shifted to major estuary systems, from the Sacramento Estuary to Chesapeake Bay. On the local and regional levels, numerous states and communities concluded that the only way to understand the division of responsibility between point sources of wastewater treatment plants and land runoff pollution, now labeled *non-point sources*, was to study an entire watershed or basin. Of course, a few watersheds, rivers, and basins had been studied during the late 1960s and early 1970s, and this was very informative as to pollutant transport in rivers. We then began to consider the concept of mass transport, rather than the concentration of a given pollutant, as a better measure of water quality impact. In this section we offer a sampling of such studies and show how the solutions evolved from pollutant removal to land use management over four decades.

Brandywine Model Project (1972–1974)

The Brandywine River, a 320-square mile watershed draining from the Piedmont Plateau in southeastern Pennsylvania into Delaware and the Christina River, has long been the subject of water resources issues, since it serves as the only source of water supply for the city of Wilmington, Delaware, as well as a number of smaller upstream communities in Pennsylvania. This small river also served as a source of power and process water for various industries during the eighteenth and nineteenth centuries, from paper and steel to gunpowder and chemicals. By the late nineteenth century the quality of water had deteriorated to critical levels, but various water experts disagreed on the sources and degree of pollution. In 1914 a liquid chlorine system was installed at the water plant, with a resulting drop in mortality from typhoid fever: from an average of 36 deaths for each of the previous 30 years, to zero.

The conflict over water quality continued between the two states through the twentieth century, with various pollution incidents fueling the political fire, and suburban growth in the Pennsylvania portion of the watershed spilling over from the Philadelphia region into former farm fields adding to the various water quality impacts. In 1972, a newly founded watershed group, the Brandywine Conservancy, initiated an effort to develop a comprehensive model of the river system, linking water quality to changing land uses within the basin. I was selected as research director, with an advisory committee comprised of Luna Leopold, U.S. Geological Survey (USGS), Ruth Patrick, Academy of Natural Sciences, and Fred Lee of the University of Wisconsin at Madison, and included participation by Notre Dame University, Penn State University, the USGS, the Stroud Research Institute, the University of Delaware, and West Chester University.

A network of continuous recording water quality stations had been established previously by CA through the Chester County Health Department and served as the framework for several years of water quality data collection and research, through 1974. The most important data set was compiled during several storm runoff periods, when the corresponding measurement of hydrographs and chemographs were observed. In addition, sampling during base flow periods gave the first documentation of phosphorus mass transport through the river network. A geographic information system was also developed, utilizing a raster-format statistical sampling design, and these land use data were correlated with the resulting mass transport flux from developed portions of the watershed. Some 14 research reports were issued in 1974, covering various aspects of land use and water quality, and the significant difference in mass transport for certain particulate-associated pollutants, especially phosphorus, between dry and wet weather flows (Figures B-3 and B-4). A model was also developed relating the dynamics of oxygen in a river system as a function of benthic conditions and the cycling of nutrients through the flowing stream.

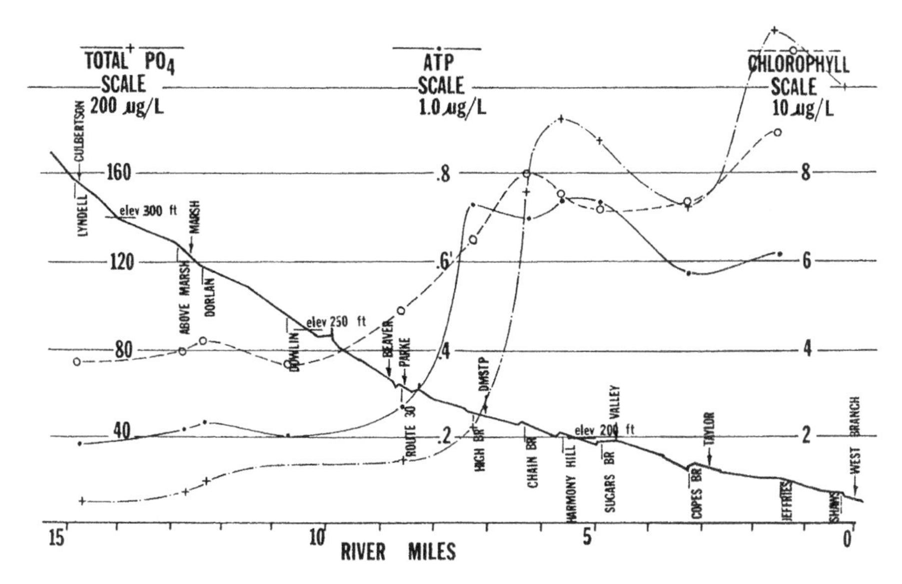

Figure B-3 Profile of total phosphorus, ATP, and chlorophyll, July 19, 1973.

Water Quality Management in the Lake Erie Basin (1975–1979)

The Lake Erie Wastewater Management Study, initiated in 1974 by the Buffalo District of the Corps of Engineers under PL 92–500, Section 108(d), focused on the input of pollutants to the lake from the surrounding drainage basin, some 23,000 square miles of which lie in the United States (Figure B-5). The primary objective of the study was to identify major sources of pollution (specifically, phosphorus) and structure a plan to restore and maintain the lake's water quality.

Early in the study it became apparent that non-point sources of phosphorus in tributary rivers accounted for a significant portion of the total mass loading to the lake, degrading the trophic level. Resource Management Associates (later to become Cahill Associates) was charged with three critical roles in this planning effort. The first was to design a comprehensive sampling program conducted by three laboratories for wet weather analysis, utilizing a network of continuous sampling stations that would develop chemograph–hydrograph data for mass transport analysis. Hundreds of storm runoff chemistry data sets were compiled for 12 tributaries over a 2-year period. In addition, the major inflow to the lake from the upper Great Lakes, the Detroit River, was sampling using a Corps of Engineers barge designed for major river systems, with multiple level samplers and velocity meters, the first time this river, with an average discharge of 192,000 ft^3/sec, had been measured and sampled in an integrated fashion.

Second, these data were analyzed to provide input for the Lake Erie model, developed by Manhattan College. Third, Cahill Associates designed and compiled a lake resource information system, with land cover/use derived from

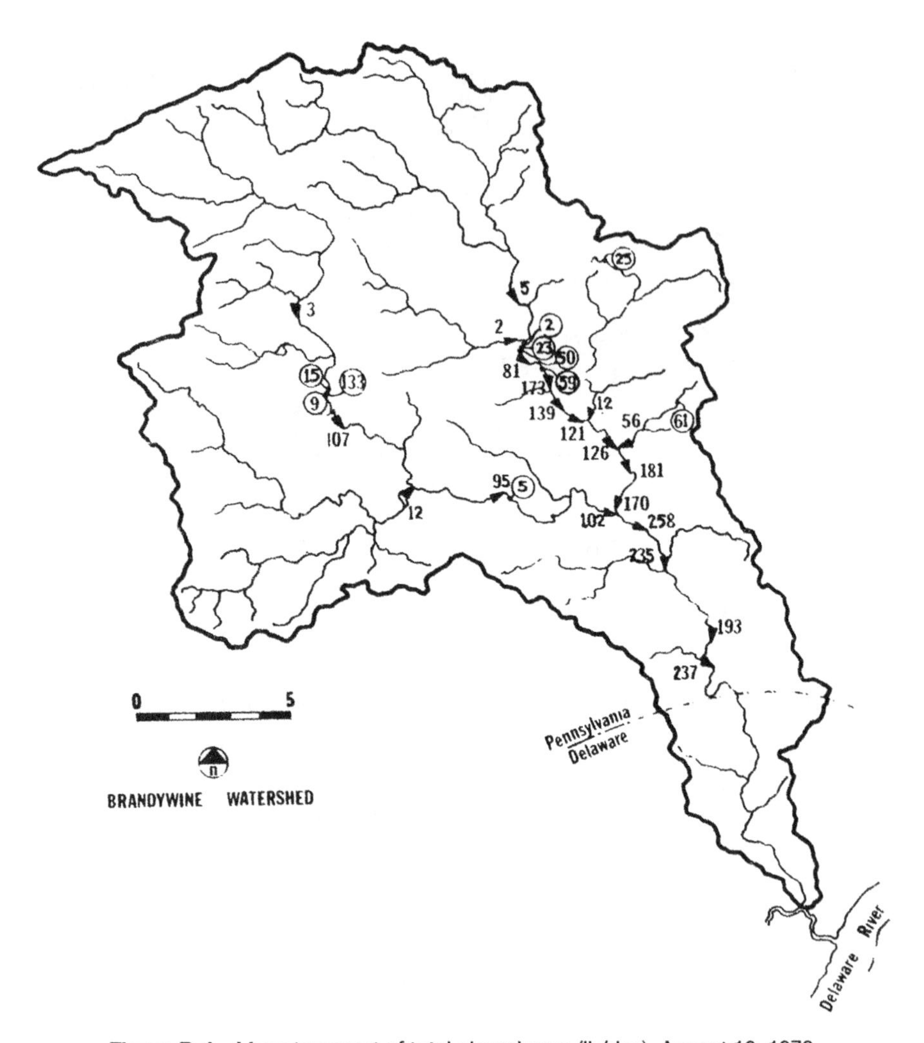

Figure B-4 Mass transport of total phosphorus (lb/day), August 13, 1973.

high-altitude aerial photography and published spatial data on soil series, geology, and political–hydrologic boundaries. The relationship between land use and water quality could be defined by the analysis of these data.

The net result of the mass transport analysis showed that NPS sources of phosphorus accounted for some 63% of the total loading to the lake, and that agricultural land use contributed a major portion of that loading. However, the input from major urban centers (Detroit, Cleveland, Toledo, Sandusky, and Buffalo) was not well measured, because they were situated directly on the lake, and virtually all of the urban stormwater runoff was conveyed to the lake by combined sewers, and the "wet weather" discharges were not sampled in the same

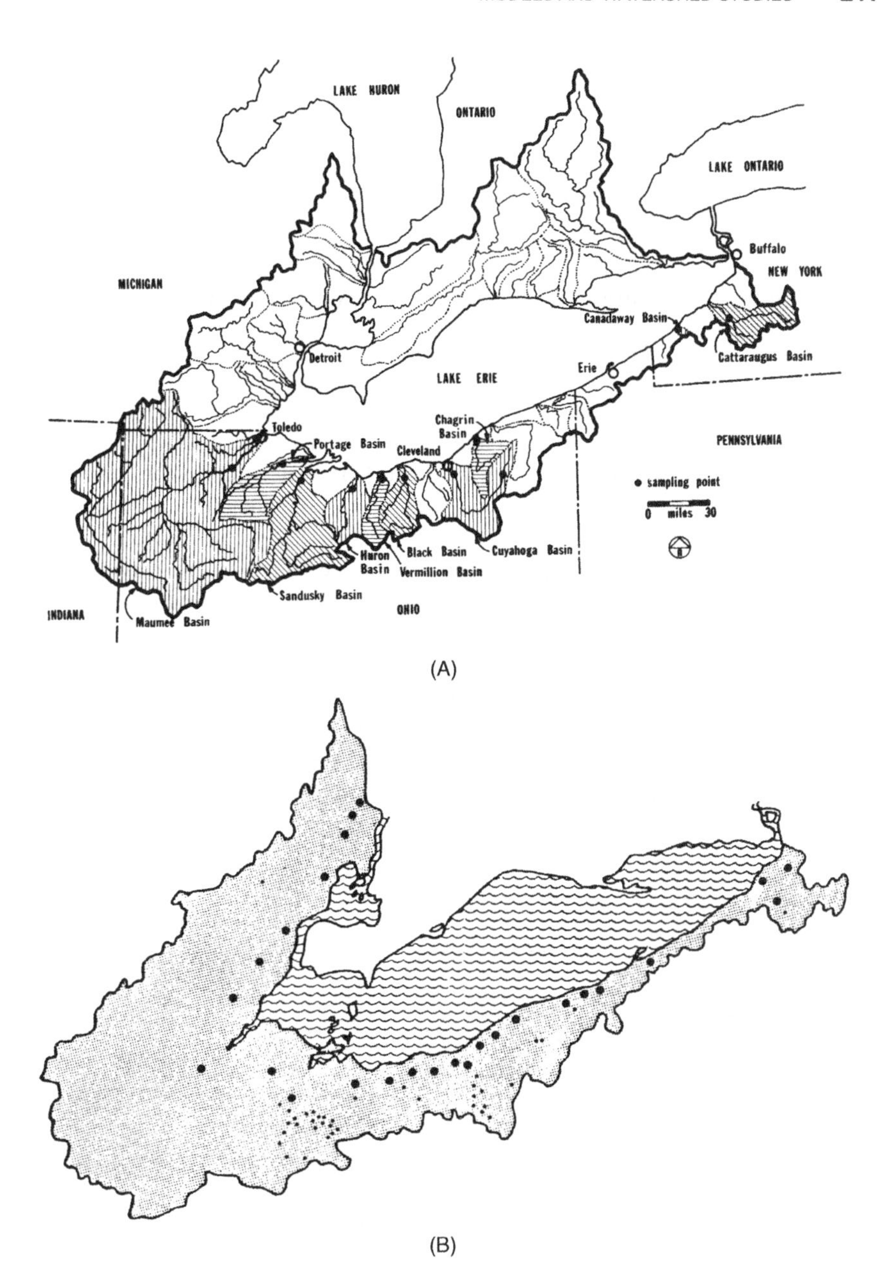

Figure B-5 Lake Erie mass transport sampling program, 1975–1979.

continuous method. The resulting management plan focused on changing agricultural practices to reduce sediment transport (no-till, minimum tillage, contour farming, and nutrient application reduction), and the then-new Detroit Regional Wastewater Treatment Facility (1 billion gallons per day) included phosphorus removal in the unit operations. However, the urban runoff contribution was not included in the plan.

Non-Point Source Model Calibration in Honey Creek (1979)

In 1977, the EPA was attempting to develop experience in the use of the NPS model designed by Hydrocomp, Inc., as well as the agricultural runoff model, also developed by Hydrocomp. The Honey Creek watershed (Figure B-6), a 187-square mile tributary of the Sandusky River in northern Ohio, tributary to Lake Erie, had been studied during the Lake Erie water quality program and a rich database of both land use characteristics as spatial data and stormwater chemistry and phosphorus transport during wet weather had been created by Cahill Associates and the Heidelberg College laboratory (directed by David Baker). With this information, the NPS model was calibrated to simulate the pollutant transport process.

The original NPS model was written for small catchments and lacked stream transport and routing subroutines in the original format, so 14 subbasins within the watershed were routed to the continuous recording gage station at Melmore. The NPS model also assumed certain flow pathways, which were different in Honey Creek, which is an intensively cultivated basin with tile underdrains creating a

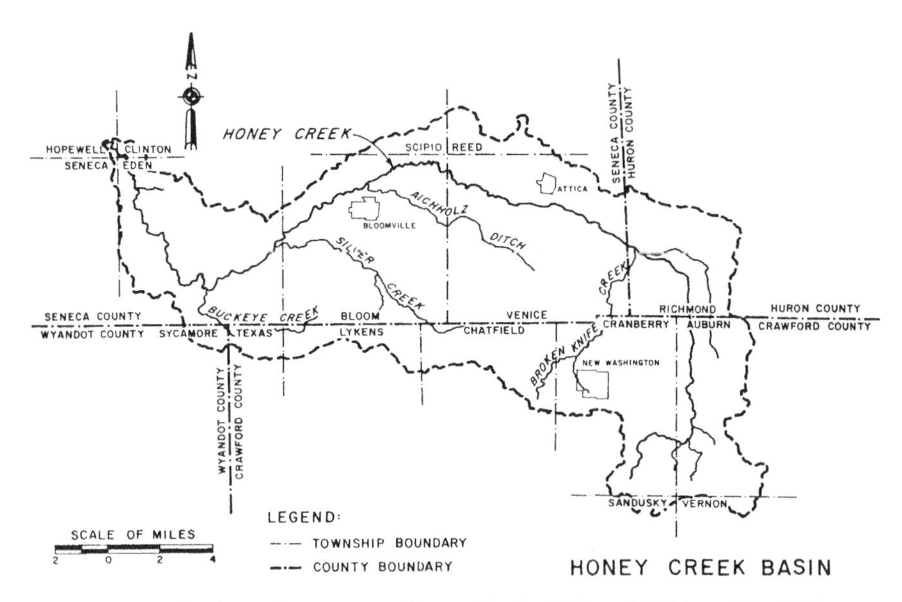

Figure B-6 Honey Creek was calibrated for the NPS and ARM models, 1977.

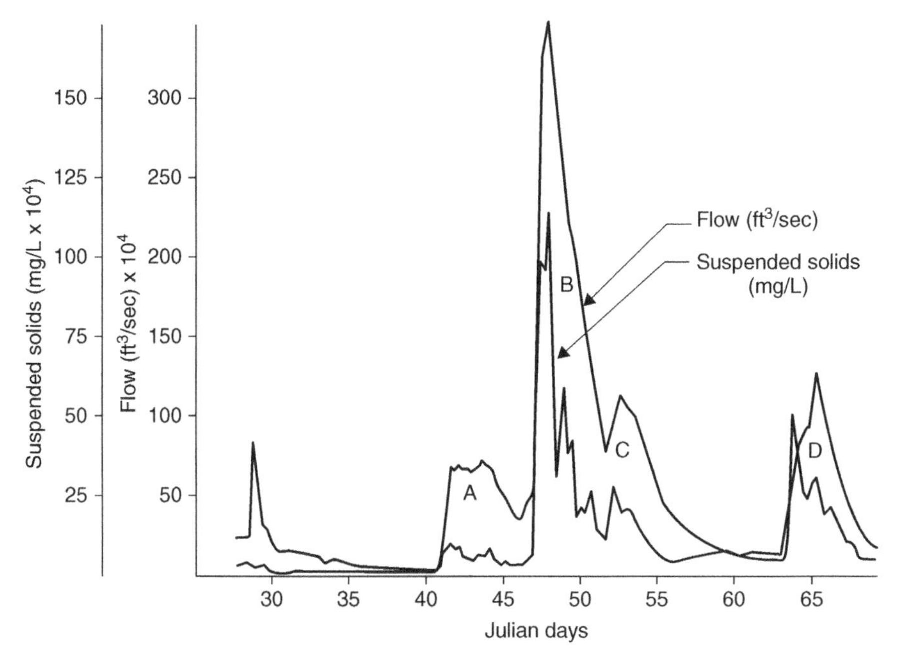

Figure B-7 Precedence of hydrograph by chemograph for multiple storms.

subsurface pathway, which limited the applicability and transferability to other watersheds. As a result, the simulated hydrographs predicted were much more extreme in magnitude than the real hydrographs, while the peaks of concentration preceded the hydrograph peak (Figure B-7).

Stormwater Management in the New Jersey Coastal Zone (1989)

The quality of water in the bays and estuaries of coastal New Jersey have been affected by pollutants contained in stormwater runoff from land development for several decades, and the associated degradation is a threat to one of the richest finfish and shellfish aquatic environments in the country. The string of barrier islands have provided shelter from Atlantic coastal currents and waves to create a rich wetlands habitat, but the occupation and paving over of both the barrier islands and mainland by increasingly larger residences and commercial parcels have created significant impervious surfaces and generated a load of nutrients from residential lawnscapes, golf courses, and other urban surfaces discharged with every rainfall. The regional hydrology has also been altered, as the natural sandy soils are sealed to increase runoff from 2 in. per year to some 43 in. on impervious surfaces.

The 2,000-square mile drainage area (Figure B-8) had been the subject of study by various investigators, and in 1989 the New Jersey Department of Environmental Protection and Energy retained Cahill Associates to prepare a plan for

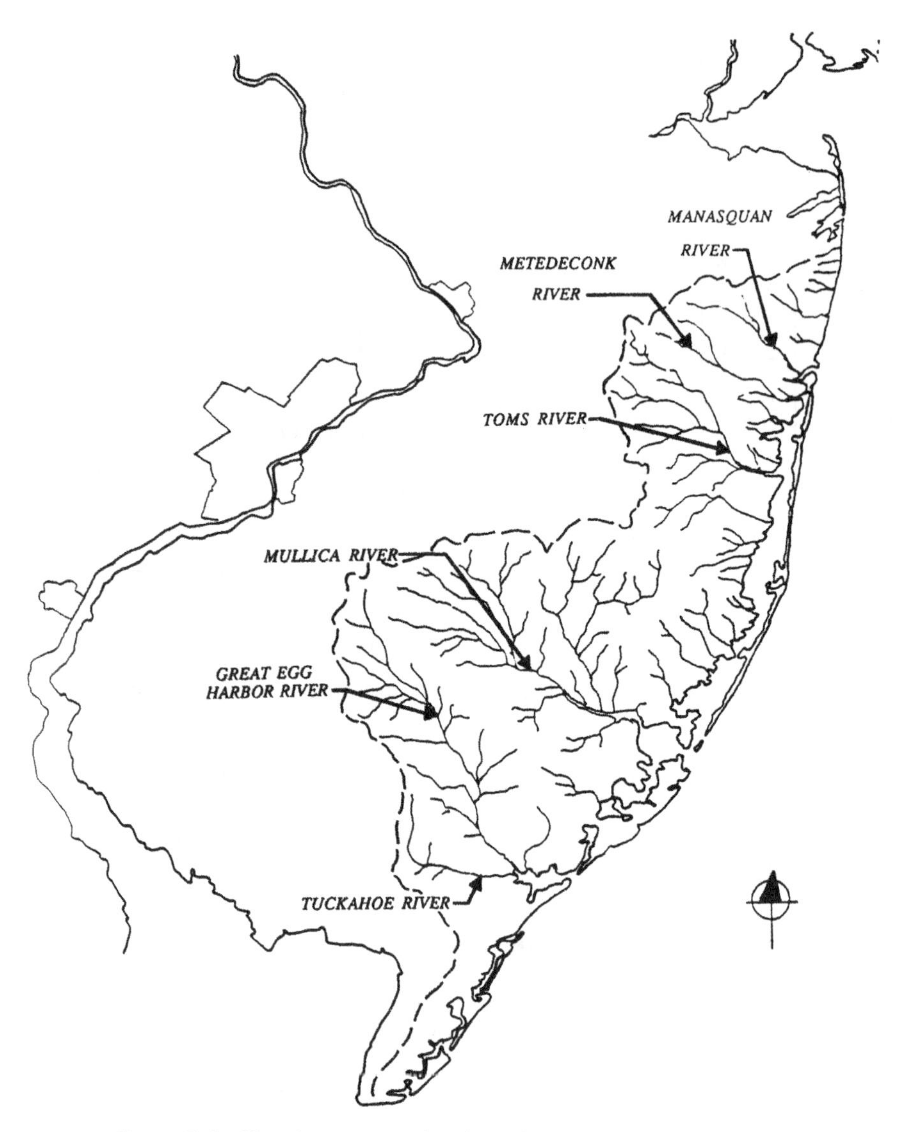

Figure B-8 New Jersey coastal drainage includes 2,000 square miles.

stormwater management in the New Jersey coastal zone, including a design manual to be used for the evaluation of proposed stormwater management systems in new land development. Cahill Associates prepared a detailed review of current practices and evaluated new stormwater management technologies for their suitability in the coastal zone. These BMPs were to be incorporated into the permit review process of the Division of Coastal Resources to reduce or prevent further pollution of coastal marine environments, and restore water quality.

Neshaminy Basin Water Resources Planning (1987–1990)

The Neshaminy Water Resources Authority, based in Bucks County, Pennsylvania (Figure B-9), was the entity responsible for the development of a multipurpose impoundment program begun in 1966, under the guidance of the Soil Conservation Service, funded under PL 566. Intended initially to function as a flood control program, with several large multipurpose impoundments designed for flood detention storage, water supply, and recreation, and a group of dry impoundments for flood control only, the system was partially constructed by 1980. One large impoundment was not yet constructed, and a water supply diversion from an adjacent river was planned as part of a supply for a then-new nuclear power plant. The agency decided to reevaluate the overall program for water supply, water quality, and flood control in light of the changing land use in the watershed from agriculture to suburban residential. It was apparent that the impact of phosphorus-laden runoff on water quality in the human-made lakes was a major problem and limited contact recreation. The impact of the river diversion on lake chemistry was also an issue.

Figure B-9 Neshaminy Basin, southeastern Pennsylvania, with dam network built under PL 566.

Cahill Associates was retained to perform a series of studies and model analyses, including flood modeling and chemical transport into the impoundments. The long-term goal was to influence land development plans within the watershed to mitigate the effects of stormwater on the human-made lakes and stream quality. Intensive meetings and conferences were held with the 15 municipalities in the basin, all of which had a different perspective on the various water resource issues.

The GVA Water Balance Model: Sustainable Watershed Management in Northern Chester County, PA (1995–2001)

The Green Valleys Association (GVA), situated in northern Chester County, Pennsylvania, has served as a technical resource for the 20 townships that are situated in five small watersheds in southeastern Pennsylvania, tributaries to the larger Schuylkill River, primary source of water supply for the city of Philadelphia and smaller communities downstream. The undeveloped land in most of these watersheds, known locally as the "green valleys" area, has resulted in very high water quality for over two centuries, but the expansion of suburban development in the late 1990s created a very strong awareness of the potential impact on both water quality and the quality of life (Figure B-10). GVA initiated an effort to plan for sustainable land and water resources, including water supply (primarily groundwater), wastewater (primarily land application without stream discharge of effluent or on-site), and stormwater, following the concept of no new increase in volume. Cahill Associates, in association

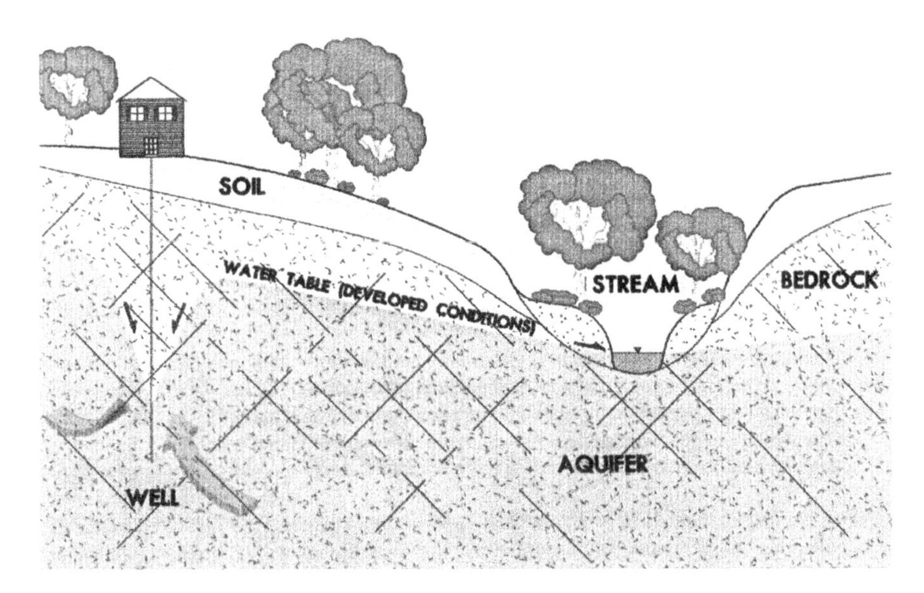

Figure B-10 Excessive groundwater withdrawals affect the stream base flow.

with the Brandywine Conservancy EMC, was retained to develop this concept, analyze the existing water cycle, identify the limits of water and land use, and formulate a land use/management program that would give the municipalities the ability to manage their land and water resources into the future. The program was supported by grants from the William Penn Foundation and the Pennsylvania Department of Conservation and Natural Resources, as well as local municipalities.

The primary goal was to sustain the essential values that distinguish these watersheds, in the face of the wave of development pressures overtaking the region. Because land use decisions directly affect water resources, and because land use decisions are made at the municipal level, a critical ingredient of this study was to work directly with the municipalities. Nine of the townships comprised the Northern Chester County Federation, and a model stormwater ordinance was developed for adoption, along with a review of planning and zoning for consistency, as well as water and sewer infrastructure systems and plans. One major concern was that the expansion of development would greatly increase groundwater withdrawals and export effluent, leading to a dewatering of the natural stream flows, with impervious surfaces creating severe ground- and surface-water quality degradation. A computer-based geographic information system was developed and used to predict future land development and water resource impacts, including a low-flow maintenance model (Figure B-11) and a dry year nitrate impact model, to ensure that unsafe nitrate levels, especially stemming from land application of wastewater effluent, did not contaminate water supplies. Technical results from these models constituted the legal basis for land management guidelines.

Valley Creek Stormwater Management Program with Reduction and Prevention of Urban Non-Point Source Pollution (1998)

The Valley Creek, Schuylkill River basin, of Chester County, is the most highly urbanized watershed in Pennsylvania with a stream classification of "exceptional value," usually applied to pristine rural watersheds. This apparent contradiction is the result of strong community interest and concern for water quality in this relatively small stream valley, which flows through Valley Forge National Park. In a sense, this watershed represents all of the small streams that lie in the path of expanding (or diffusing) urban regions which have been altered by the changing patterns of transportation, land use, and economic development. As new highways have reached out into the countryside surrounding every urban center, carrying both residents and employment into valleys that were farmland some 30 years ago, they have covered over the land with impervious structures and surfaces. In this watershed, the impervious cover was some 20% of the total watershed in 1998, with about 65% comprised of roadway surfaces and parking lots, serving the automobile. This produced a significant increase in runoff and the pollution generated from these impervious (and pervious) surfaces.

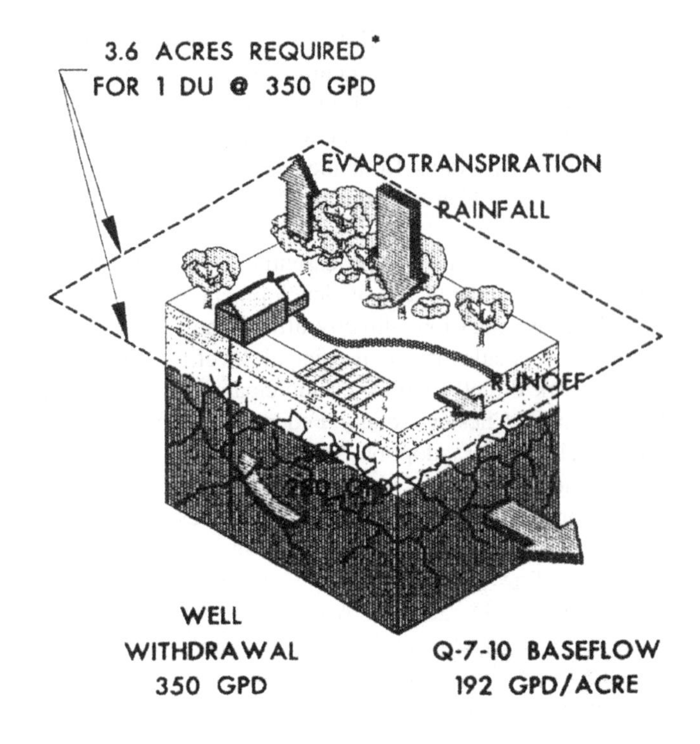

MAX. ALLOWABLE BASEFLOW REDUCTION:

10% OF 192 GPD = 19.2 GPD

* Assumes Stormwater Recharge

Figure B-11 Low-flow groundwater maintenance model was based on nitrate levels.

The flooding impacts on the riparian channel could be estimated by the increase in bankfull flow from once in 18 months to once a month, carving the stream and degrading fisheries habitat. While still supporting an excellent trout population (with restocking), water quality has continued to deteriorate as land uses change. A program was prepared for implementation by local governments for the use of both nonstructural and structural stormwater management measures, but was met with resistance by development interests, resulting in a series of litigations in county and state courts.

The various solutions were supported by the County Conservation District, the Valley Forge National Park, Trout Unlimited, conservation groups, and the municipalities, and the BMP effectiveness was monitored with the use of rapid bioassessment protocol to evaluate the suitability of methods. The watershed was subdivided into 40 subareas with over 120 detention basins, and their long-term impact on watershed hydrology was evaluated and modeled.

Figure B-12 The University of Virginia campus was evaluated for the application of LID.

Strategic Plan for Water Resources Management at the University of Virginia (1989)

At the University of Virginia, Charlottesville, Virginia (Figure B-12), Cahill Associates and Andropogon Associates performed a stormwater management and land use study of the entire 1,300-acre campus. The goal was to manage and restore the natural resources of both land and water, and to plan for the careful integration of new facilities while preserving the quality of the campus. A geographic information system served as the resource data manager and analysis framework, with various hydrologic and hydraulic models employed to select the most beneficial technologies for each catchment. The initial analysis focused on the gardens designed by Thomas Jefferson in 1822, one of the most historic landscapes in the country. Drainage and landscape solutions were proposed to preserve and protect these vital features. Also included in the recommendations was the "daylighting" of the small stream that was buried underneath portions of the campus, a design that was implemented several years later.

LEED V2 Standards Development (1999)

The Leadership in Environment and Energy Design (LEED) program had its beginnings in the mid-1990s, with an effort by a mix of architects, engineers, manufacturers, and other parties with an interest in formulating a comprehensive set of guidelines on how to design and build more environmentally and energy-efficient buildings and sites. By 1999, they held a series of brainstorming workshops with a broad mix of experts in various fields. The meetings were held at the Rockefeller mansion along the Hudson River north of New York City and were sponsored by the Rockefeller Foundation. I was asked to participate in these two weekend "retreats," during which time the participants developed the set of "credits" that became the foundation for the current program. In particular, the Site/Water Credit 5, Reduce Site Disturbance, Site/Water Credit 6, Stormwater Management, and Site/Water Credit 7, Water Efficient Landscaping, were my primary contribution. The reduction of runoff volume by infiltration technologies was a constant theme and found its way into the system, with some modification and continued inclusion of detention concepts.

The LEED program developed dramatically over the next decade and has now become the "standard" for green building design technologies. However, the concept proposed by Arcadia is a step beyond the original LEED concepts, with zero net water and energy demands as the proposed standards. For many existing retrofit designs, these goals will require independent water and energy systems that effectively go "off the power grid" and disconnect the public plumbing, steps that will meet with resistance in most urban communities.

Rockaway River Sustainable Watershed Management Plan (1999)

A century ago, as New Jersey suburban communities that were part of the New York metropolitan region and lined the west side of the Hudson River developed into independent cities and industrial/commercial centers of their own, they sought a source of water supply. To the immediate west lay a number of small watersheds, situated in a physiographic region known as the "Highlands" (Figure B-13), which yielded a pristine supply of water from largely wooded

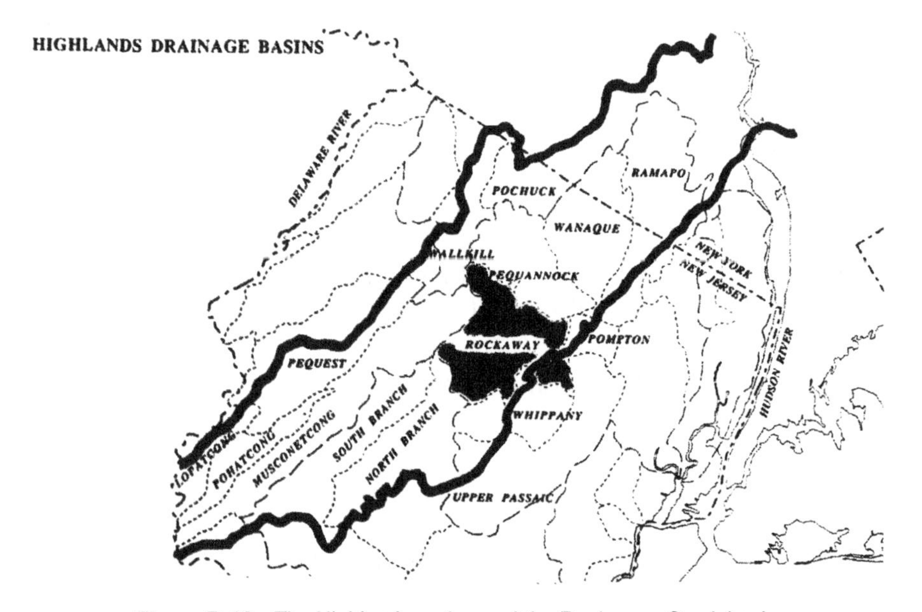

Figure B-13 The Highlands region and the Rockaway Creek basin.

watersheds. Jersey City built a large reservoir in one such basin, the Rockaway River, and currently withdraws some 42 million gallons per day as a raw water source, which seemed suitable in a basin with a daily average rainfall of 291 million gallons per day. The river valley is also characterized as containing a valley fill of glacially deposited material, forming an excellent groundwater reservoir.

As the region has grown with its own suburban land use, the question of how to accommodate this changing land cover and the related pollution load generated by development has become an important issue. The 14 municipalities that comprised a loosely affiliated watershed cabinet concluded that it needed to develop a "sustainable watershed management plan" to protect the quality (and quantity) of water resources within the basin for local use, independent of the use of basin outflow by seemingly remote communities, many now aging, to the east. Since the towns and townships that comprise the watershed residents are quite different in character and form, each with its own set of zoning, development, and land use regulations, the only common interest was the watershed. They developed a set of goals (Figure B-14) that it is hoped will form the basis of all future land use, and a strategy of stormwater management that required infiltration, to assure that the critical aquifer system would be maintained into the future (Figure B-15).

B.5 DESIGN AND CONSTRUCTION PROJECTS

Environmental Master Plan for the Main Campus and Mason Farm, University of North Carolina, Chapel Hill, NC (2002–2004)

As a supplement to the campus master plan, Cahill Associates and Andropogon Associates prepared an environmental master plan, intended to study the impact of proposed development and redevelopment, funded under a state grant of $2 billion. The team examined the campus from the perspective of land and water resources, and focused on achieving a balance between human uses and the natural resources of the 729-acre campus, within the sustainable limits of the natural system. The plan examined the land use from a watershed perspective, with a focus on impervious areas, historic and buried streams, and the natural hydrologic balance on the campus (Figure B-16).

Chapel Hill is recognized for its harmonious blend of mountain and coastal vegetation, from which an extraordinary garden community has emerged. Beginning with the first building constructed in 1793, the central campus of the university incorporated the significant groves of trees into its architectural beauty. The intent of the plan was to reconnect the north and south campus areas with their natural and vegetative systems.

Five creeks contribute to the drainage on both central campus and Mason Farm, including Meeting of the Waters Creek, a historic stream that flows beneath much of the campus, including the football stadium and other structures. The study identified sustainable strategies to mitigate the increased flow and downstream

GOALS OF THE ROCKAWAY RIVER SUSTAINABLE WATERSHED MANAGEMENT PLAN

Sustain and Restore the Quantity and Quality of Streams and Groundwater

- Maintain stream base flows in all perennial streams – Don't let the streams go dry.
- Protect aquatic communities, habitats, stream channels, and corridors.
- Protect intermittent channels as flow pathways.
- Maintain wetlands and related vegetative systems.
- Protect the quantity and quality of existing and future wells.
- Reduce nonpoint source pollution of ground and surface waters.

Reduce Existing Flooding and Prevent Increases in Future Flooding

- Protect the riparian corridor from filling, vegetation loss and structures.
- Remove old fill from the regulated flood plain.
- Prohibit placement or storage of pollutants within the floodplain.
- Remove existing structures from the regulated floodplain where feasible.
- Retrofit existing stormwater infrastructure to increase efficiency.
- Prevent increases in stormwater runoff volume from new development and reduce where possible excessive runoff from existing development.

Protect Lake Systems by Better Land Management

- Develop improved guidelines for land disturbance and fertilization practices.
- Reduce/prevent wastewater discharges to lake systems.

Figure B-14 Land use and water quality goals of the Rockaway basin.

effects of increased impervious surfaces, including the use of porous pavement to restore groundwater recharge, restoration of forested valley slopes, preservation of existing forested stream lowlands and daylighting of buried streams, in order to restore the hydrologic balance for this campus. The university created a GIS database to facilitate this planning. The plan received an Excellence in Planning Award in 2005 from the Society for College and University Planning, AIA.

Prior to and as part of the campus planning process, Cahill Associates prepared the Hydrologic and Hydraulic Model Analysis of Potential Flooding Conditions (March 2004) for a portion of the campus, followed by the Stormwater Management Plan (October 2004), prepared in conjunction with Andropogon Associates

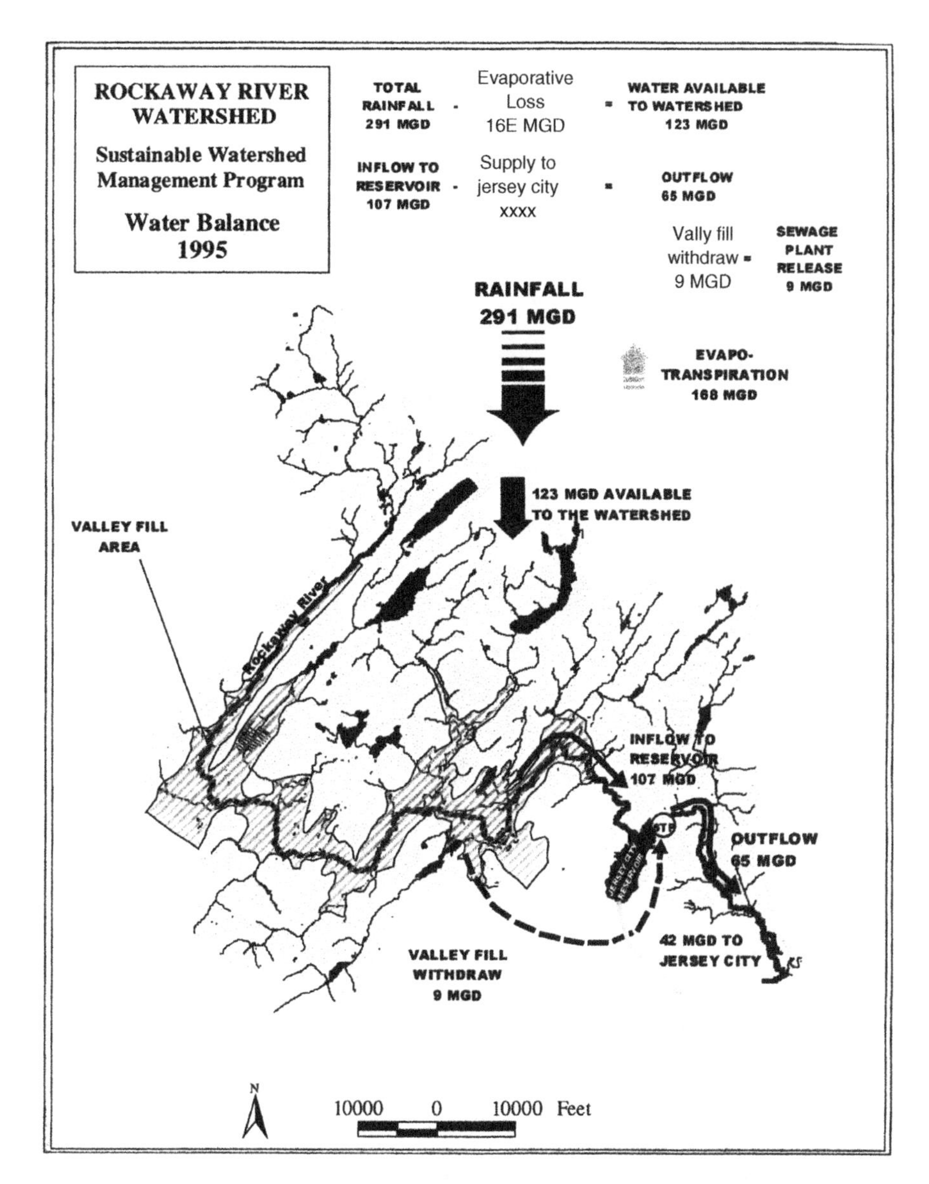

Figure B-15 Water balance in the Rockaway River basin.

and The Rose Group. Because of the richness of the GIS database (developed by University of North Carolina Facilities Department), a very interesting analysis of the potential application and cost benefit was possible for the various measures considered possible in the design program. Figure B-17 summarizes this information, and was an important part of the decision-making process as campus construction moved forward.

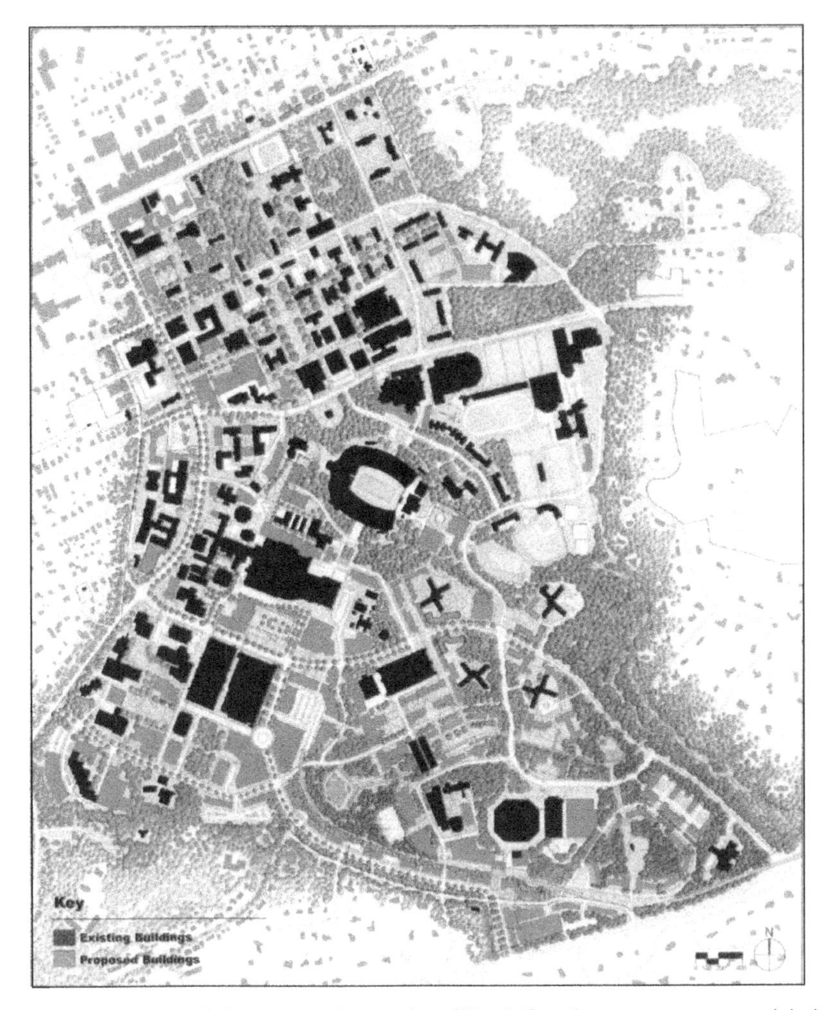

Figure B-16 The highly impervious University of North Carolina campus was modeled for LID measures.

Green Infrastructure Planning and Design for East Liberty Neighborhood, Pittsburgh, PA

A major redevelopment effort is taking place in Pittsburgh's East Liberty neighborhood (Figure B-18), begun in 2007–2008 by Cahill Associates and Veridian Landscape Architects and continued by CH2M HILL and Veridian Landscape Architects (2009 to the Present). East Liberty, historically a fairly disadvantaged community, is undergoing a revival that contains some massive demolition and reconstruction efforts, transforming this area into a thriving successful retail and residential focal point. An important component of the redevelopment and master

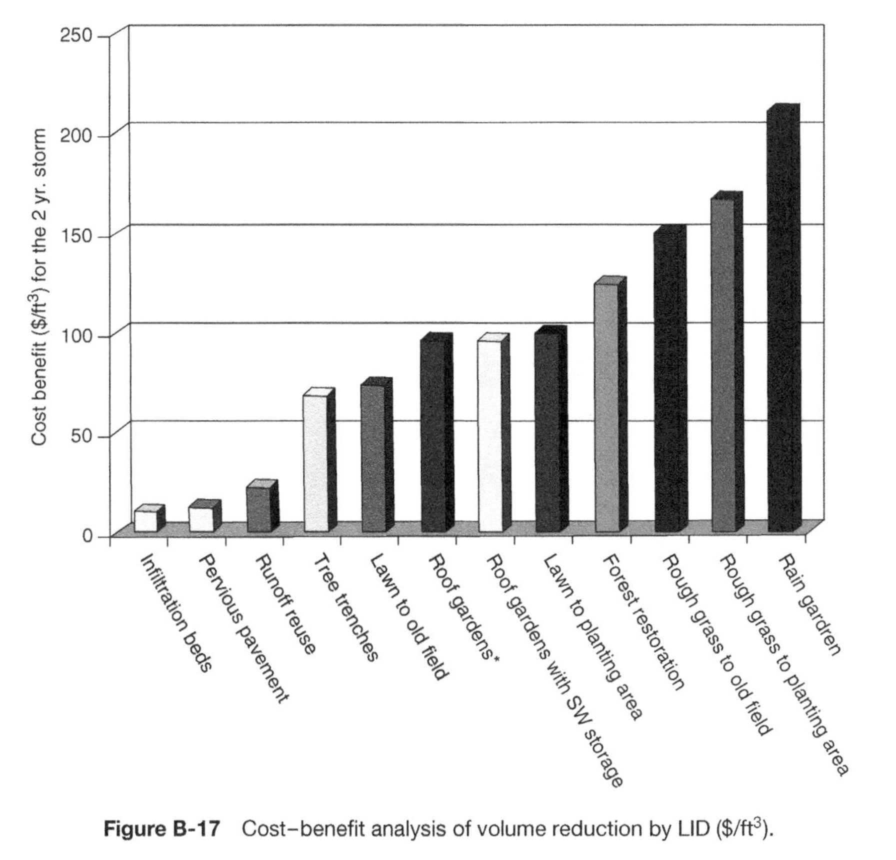

Figure B-17 Cost–benefit analysis of volume reduction by LID ($/ft^3).

Figure B-18 East Liberty residences have small lots and sloping streets.

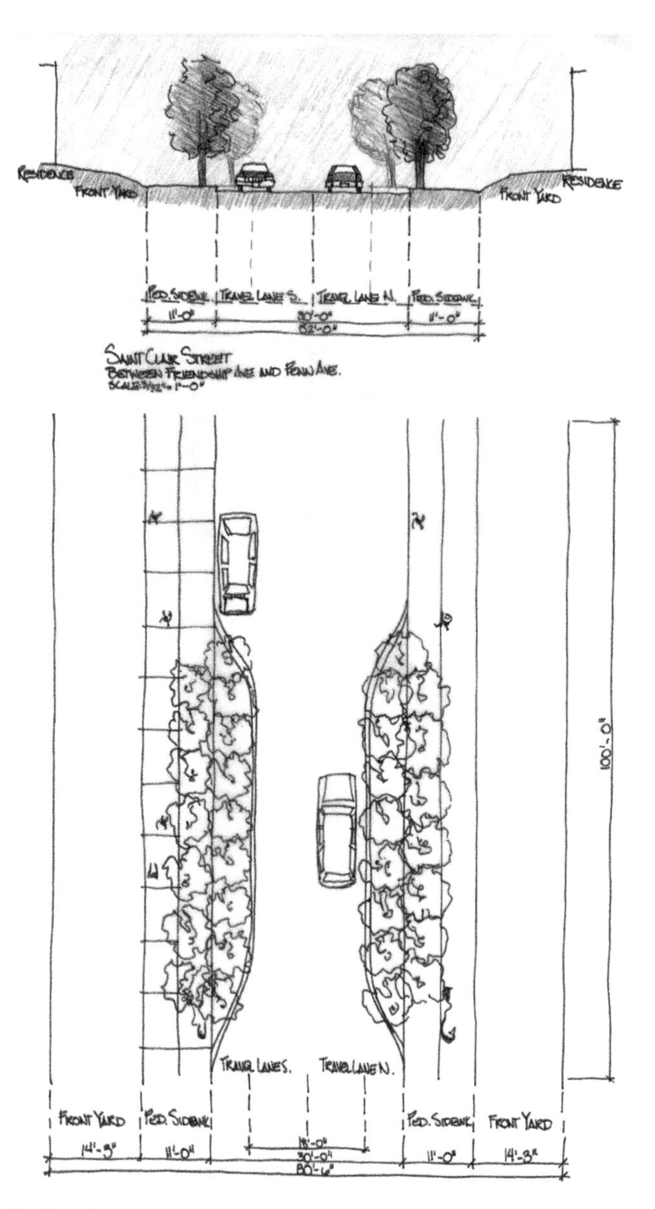

Figure B-19 Green street design combines vegetation and porous pavements.

planning efforts is the application of sustainable design concepts to the new development, incorporating "green" elements that will reduce the impact of stormwater runoff on the sewer system, reduce energy demands, provide a cleaner, healthier environment and greatly improve the quality of life for both existing and new residents (Figure B-19).

Much of the community's sewer system, roadway and streets, pedestrian sidewalks, and other infrastructure elements are planned for reconstruction and offer exciting opportunities for immediate greening of the urban fabric. The residential community in East Liberty provides such an opportunity, with small "backyard BMPs," identified as attainable solutions to the greening goal. A combination of *green infrastructure* measures will reduce the total impact of runoff to the combined sewers to a level that will reduce the frequency of overflows during extreme rainfall events, which combined with efforts to reduce surface pollutant inputs should help in the restoration of water quality in the receiving rivers.

Allegheny Riverfront Vision Plan (2010)

The city of Pittsburgh, Pennsylvania, has undergone a major transformation during the past four decades, as river-based industrial production facilities for steel, aluminum, iron, glass, and related water-transported finished materials have moved from the region and left behind a waterfront that contains some residential and commercial activity, but by and large is in need of redevelopment. As the Urban Redevelopment Authority of Pittsburgh has considered this issue, they retained a consulting team led by Perkins Eastman, Architects and Planners, and including CH2M HILL and Veridian Associates, Landscape Architects. This team was charged with formulating a "Vision Plan" that included the best elements of land redevelopment, such as LID, and would lead the city to a livable, economically viable, and environmentally sustainable future, creating a vision that would ultimately extend throughout the city from the waterfront of the Allegheny River (Figure B-20).

The plan envisioned both the restoration of existing residential communities and "greening" of commercial and community parcels, as well as setting specific goals for both stormwater management and landscape design. The city is burdened with combined sewers, and during heavy rainfall some 52 CSOs discharge to the river along this waterfront, severely polluting the river and preventing a comprehensive redevelopment program. The restoration of hydrology includes the planting of trees throughout the 1,600-acre study, ultimately achieving a

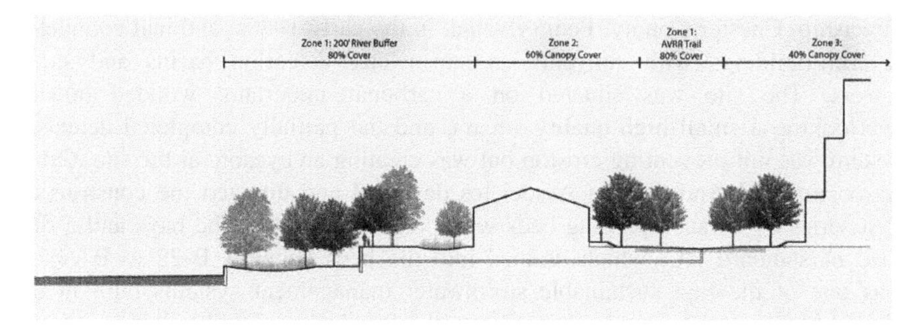

Figure B-20 The Allegheny Riverfront Vision integrates vegetation with structures.

Figure B-21 Green buffer will form the river edge as a pedestrian corridor.

40% canopy on average, with the river "greenbelt" edge achieving the greatest percentage (Figure B-21).

A key component to this vision is to get private developers to participate in the program, effectively reducing any public cost associated with sustainable redevelopment. A local rail line is also a critical part of the plan, with the right-of-way turned into a green transit corridor. This greenway will be a part of the stormwater volume-reduction plan, which includes porous streets, sidewalks, rain gardens, vegetated roofs, and other solutions to keep all runoff from entering the existing combined sewer system.

Shared Medical Systems (now Siemans) Sustainable Stormwater Management System (1982)

The SMS Corporation was building an office complex in East Whiteland Township, Chester County, Pennsylvania, in the early 1980s, and had completed several buildings with conventional stormwater detention basins and storm sewers. The site was situated on a carbonate-underlain, wooded hillside overlooking a small high-quality stream, and the partially completed detention system was not preventing erosion but was creating an eyesore at the site. Cahill Associates and Andropogon Associates designed and directed the construction of a series of terraced parking beds with porous AC within the bays and a ring road of standard AC, which drained into the beds (Figures B-22 to B-24). It was one of the first sustainable stormwater management systems built in the United States, and served to demonstrate the feasibility of infiltration as a better solution for stormwater runoff volume management. It has been in operation

Figure B-22 Geotextile separated a storage bed from soil in early designs.

Figure B-23 Unpaved edge becomes a drain in event of pavement clogging.

for three decades and continues to function. Anticipated issues of ice damage, cracking, settlement, and deterioration have never materialized.

As a point of interest, the surface seems to have a smooth and stable appearance, and no cracking or weathering is apparent. Cleaning has been on a random basis and has diminished over time as site maintenance has changed. Other portions of the site, paved with conventional AC, have required resurfacing twice since 1982. The carbonate geology has also created sinkholes on other portions of the site around inlets set in impervious pavements.

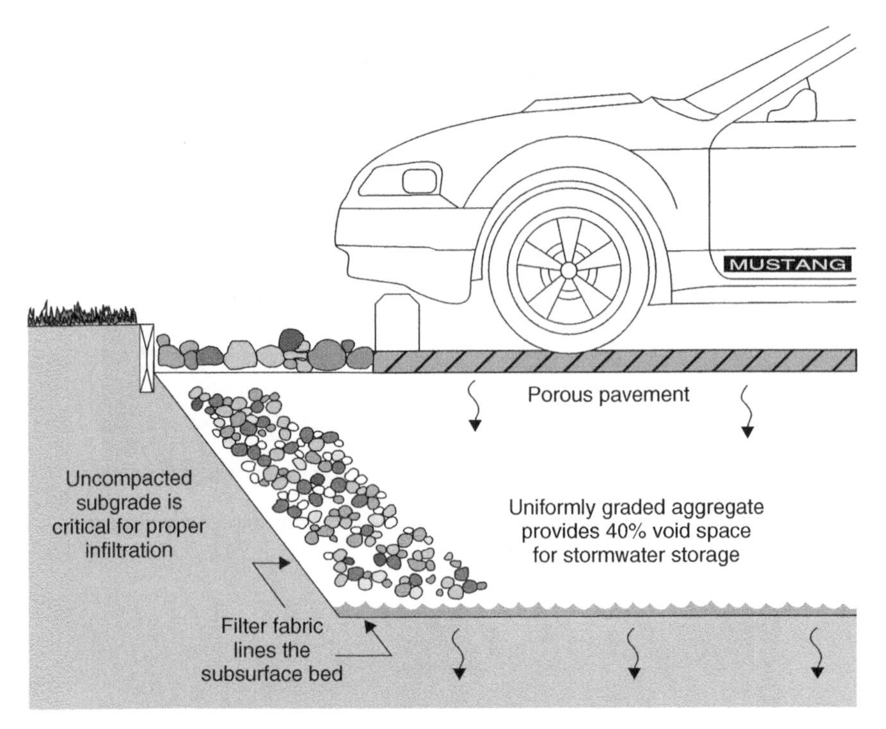

Figure B-24 Surface connection to infiltration bed.

Morris Arboretum Stormwater Management System and Entrance Driveway (1984)

The Morris Arboretum of the University of Pennsylvania, situated in northwestern Philadelphia, Pennsylvania, serves as the repository of all tree species in the state, and is a very important environmental resource and educational institution. It is also one of the major public garden spaces in the region and serves as a model of sustainability in the urban center.

In 1983, it was determined that public parking was inadequate on the site, and in addition a new entrance drive was required. A strong sense of resource protection motivated the institution to select an innovative design proposed by Cahill Associates and Andropogon Associates, using porous AC pavement in the parking bays of the lots (Figure B-25), and placing a bed of uniformly graded crushed stone beneath the entire system, including a center drive with standard AC pavement. In addition, the entrance driveway was designed with lateral infiltration trenches along the slope so that rainfall was conveyed into open meadows from inlets (Figure B-26). The final site design prevented runoff, and in fact the trees surrounding the parking lot have grown to be larger and healthier specimens than any others, irrigated by the infiltration beds.

Figure B-25 Porous pavement was used only in parking bays.

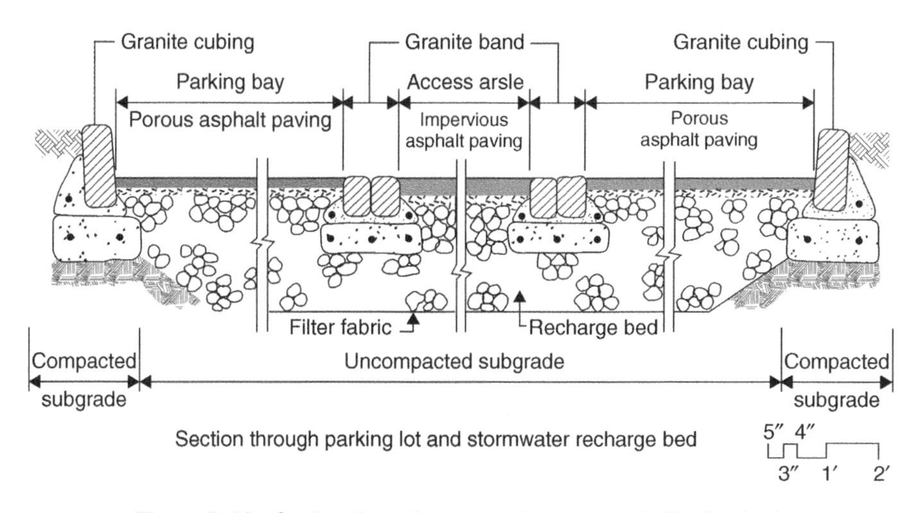

Figure B-26 Section through pavements and stone infiltration bed.

Campus Master Plan, DuPont Barley Mills Office Center, New Castle County, DE (1985)

Cahill Associates and Andropogon Associates were retained by DuPont to develop a sustainable plan for a new office complex that would serve a number of company subsidiaries, including the worldwide headquarters for agricultural chemicals. The land development program was under way when they were retained, but they were able to design a new parking lot system for this part of the site, using porous AC over a stone recharge bed, and including the rooftop

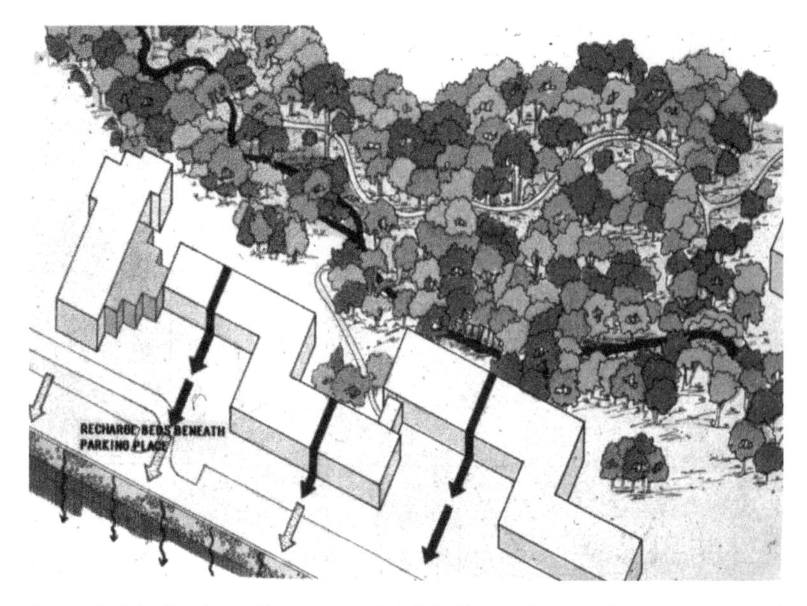

Figure B-27 Roof runoff conveyed to infiltration beds, wooded area preserved.

runoff from several large office buildings (Figure B-27). Review and approval by county government was cautious at the time, based on a lack of experience with recharge designs. However, six months after construction, hurricane Gloria dropped 6 in. of rain on the site within 16 hours (Figure B-28). The performance of the system and the fact that no overflows occurred and no increase in stream flow resulted provided an excellent demonstration of both the use of porous pavements and the infiltration of rainfall as the best way to prevent runoff impacts (Figure B-29).

Clinical Laboratory Complex, SmithKline Beecham (now GlaxoSmithKline), Valley Forge Business Center, Montgomery County, PA (1987)

Shortly after completing an evaluation of SmithKline Beecham's animal health products headquarters complex at Applebrook, Chester County, Cahill Associates and Andropogon Associates were asked to design a new site to serve a $17 million, 135,000-square foot clinical laboratory. Water resource concerns posed the most serious constraint, given the location and sensitivity of the surrounding woodlands. Cahill designed a terraced set of parking bays with porous AC pavement, underlain with stone infiltration beds, which could infiltrate the 2-year frequency rainfall from the full site and mitigate greater storms that would overflow to the nearby stream (Figure B-30). This allowed Andropogon to save a mature oak–beech forest and create an enclave for the office–laboratory complex. Included were re-created wetlands and innovative

Figure B-28 Porous pavement during hurricane Gloria, September 21, 1985.

Figure B-29 Both geotextile and plastic pipe were manufactured by DuPont.

vegetation management techniques. This project received the Planning Merit Award from the Montgomery County Planning Commission in 1990.

Urban Stormwater Management for the Penn-Partnership K–8 Public School, Philadelphia, PA (2003)

At the Penn-Partnership K–8 public school, Cahill Associates designed several innovative storm water management techniques to limit flooding, increase groundwater recharge, demonstrate storm water cleansing, and provide curriculum for the students by creating educational opportunities associated with these projects. Located in West Philadelphia, in an area with combined storm and sanitary sewers and little green space, the need was acute.

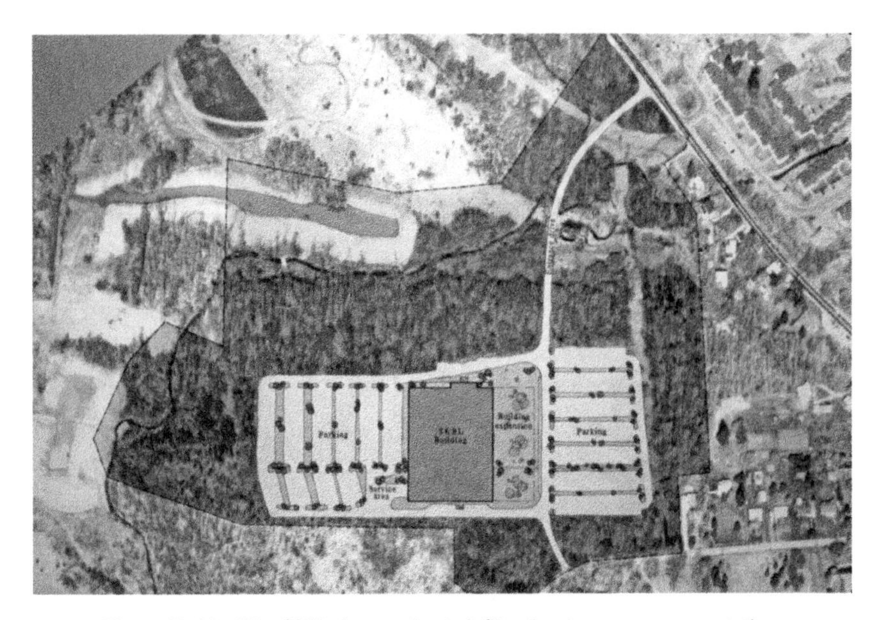

Figure B-30 The SKB site used only infiltration to preserve vegetation.

Three separate systems were designed as part of an overall site plan to capture nearly all of the roof runoff from the new buildings, and most of the site runoff, significantly reducing or eliminating the runoff going into the combined storm and sanitary sewers during storm events (Figure B-31). A large gravel stormwater storage and infiltration bed was placed under the school's turf athletic field to retain water from the new building roof drains. A porous pavement playground absorbs rainfall that falls directly on the surface (Figure B-32).

A rain garden captures roof runoff during storm events, filling a shallow basin with gently sloping edges to a maximum depth of 15 in. Water in the rain garden gradually infiltrates and disappears totally during dry periods, mimicking ephemeral pools. A large wooden viewing platform makes it possible for teachers to bring their classes outside, while boulders of natural stone allow individual students to step among the cattails and experience the pond firsthand. This little area provides habitat for a rich variety of insects, toads, birds, and other creatures that inhabit ephemeral ponds, not found in the community.

Urban Stormwater Restoration for the Washington National Cathedral (2001)

With the construction of the Washington National Cathedral at the top of a wooded hill in Georgetown, DC, the amount of runoff generated from the site was greatly increased (Figure B-33). Stormwater runoff was piped and concentrated into a ravine in the historic Olmstead Woods, causing severe erosion. By concentrating the flow, the soil mantle was denied sufficient water to keep the

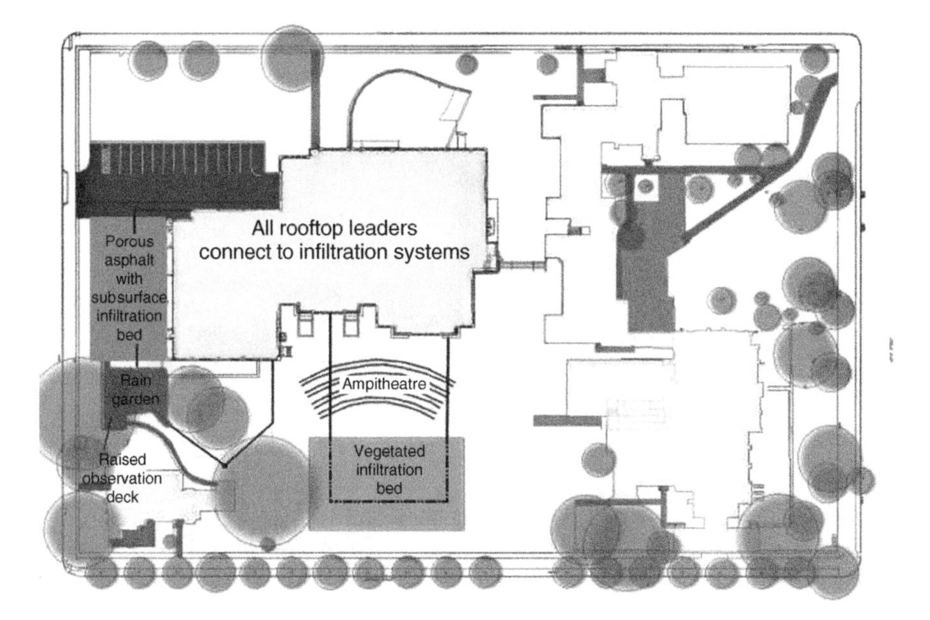

Figure B-31 The school was redeveloped in the urban center.

woodlands healthy, and the historic trees were dying. To remedy this stormwater problem and restore moisture to the woodlands, Cahill Associates and Andropogon Associates designed a series of infiltration trenches, beds, and systems designed to return the water to the soil and prevent erosion (Figure B-34). An infiltration trench adjacent to the sidewalk on the top of the hill was constructed to retain stormwater volume. The eroded "ravines" were restored with check dams. This eco-engineering approach is now endorsed by the cathedral, which holds tours for watershed groups and officials.

The cathedral continued this philosophy of "stormwater for the natural system" when the Amphitheater Restoration Project was undertaken in the summer and fall of 2006. Working together with Michael Vergason Landscape Architects, Cahill Associates designed a site with minimal grading and a design to return water to the soils and vegetation. Shallow infiltration trenches at the base of each new seat wall infiltrate stormwater into the soils above the historic Olmsted Woods. The amphitheater's stormwater management system effectively recharges the groundwater, helps maintain healthy woodlands, and reduces erosion and degradation in a rare spiritual and natural sanctuary in the busy DC area.

Ford Motor Co., River Rouge Plant, Stormwater Engineering (2001–2003)

Through a project called Heritage 2000, Ford Motor Company has become an industrial leader with a new commitment to the environment at the River

(A)

(B)

Figure B-32 The playground is porous AC and the playfield has an infiltration bed.

Rouge plant complex, Dearborn, Michigan. During a $2 billion plant renovation, with environmental leadership by William McDonough & Partners, Cahill Associates was brought in to provide sustainable stormwater management planning and design. The first of many improvements at the River Rouge plant included new Mustang vehicle storage and staging parking lots. Cahill Associates designed a system using porous pavement, subsurface storage, and water quality swales to capture and treat stormwater runoff (Figure B-35). Similar

Figure B-33 The National Cathedral sits on a wooded hilltop in Georgetown, DC.

systems have been employed throughout the facility, including vegetated roof covers on the new 1 million-square foot assembly plant, runoff storage, and an open surface constructed wetland for water quality. Together the team integrated stormwater management with landscaping and building layout and design, for an environmental vision of the future.

The Village at Springbrook Farm Residential Development, Hershey, PA (2002–2003)

This project was a high-density multifamily residential development that incorporated LID principles and sustainable stormwater management methods to reduce the impact on the larger watershed, the Susquehanna River basin, a primary source of the Chesapeake Bay. While the preexisting land use was cultivation, the regional land use surrounding the community of Hershey is rapidly changing

(A)

(B)

Figure B-34 Infiltration bed build in the wooded hillside with plastic storage units.

to suburban residential. The Village at Springbrook Farm is a 259-unit development with 149 townhouses, 96 quads, and 17 single-family homes, advertised as a LID community in Campbelltown, near Hershey, Pennsylvania. The site design approach was to keep the rainfall as close to the buildings as possible and infiltrate. The site overlaid a carbonate aquifer, and the issue of sinkhole formation influenced the placement of all stormwater infiltration beds, with potential locations carefully avoided. Some 124 infiltration beds were situated throughout the site, under porous pavements, driveways, sidewalks, and open spaces (Figures B-36 and B-37). Since the regional water supply is drawn from the limestone aquifers, this design represented recycling of all incident rainfall for withdrawals by wells. For the single-family residential units, rain gardens were used for rooftop infiltration (Figure B-38). This design met with initial resistance from homebuyers until the vegetation became an aesthetic benefit to the units.

Figure B-35 Ford porous pavement discharges to wetland trenches that drain to the Rouge River, assuming no infiltration.

Ford Amazon Plant (2003)

Following successful redevelopment of the River Rouge plant in Dearborn, Michigan, during 2001–2002, Ford considered other plant locations around the globe where the principles of sustainable or "green" design might be suitable. The company had long wished to develop a truck and auto manufacturing plant in South America, to gain a greater share of that rapidly growing market. The Brazilian government had a site available in the northern state of Bahia, in the Atlantic rain forest, where a prior developed had cleared and stripped almost 2 miles (1,236 acres) of forest for a manufacturing facility, and then changed its mind, leaving a totally barren and lifeless landscape generating tremendous erosion and runoff (Figure B-39). Ford decided to send the same team of consultants, led by Bill McDonough, to Brazil, to formulate a concept design with the local term of consultants and designers, who would implement and construct the building complex. A critical requirement by the Brazilian government was that the final site plan would include restoration of the original rain forest ecology, a challenge that had never been accomplished with a new land development project in the country.

Through a series of working sessions and field trips, meetings with local experts in soil sciences, ecology, and vegetation, including faculty at the University of Salvador, Bahia, a basic strategy was formulated by Cahill Associates for site restoration and hydrologic cycle capture and reuse, to eliminate any runoff. Included in the water planning was a wetland wastewater treatment system, in-plant reduction of water use and rooftop capture–reuse, porous pavements in all parking areas, and the use of waste organic solids to build a new soil mantle for planting beds of native trees. It was estimated that the rain forest restoration would require some 20 years but would take place over the major portion of the site. The following year, Ford experienced a significant fiscal blow when tires

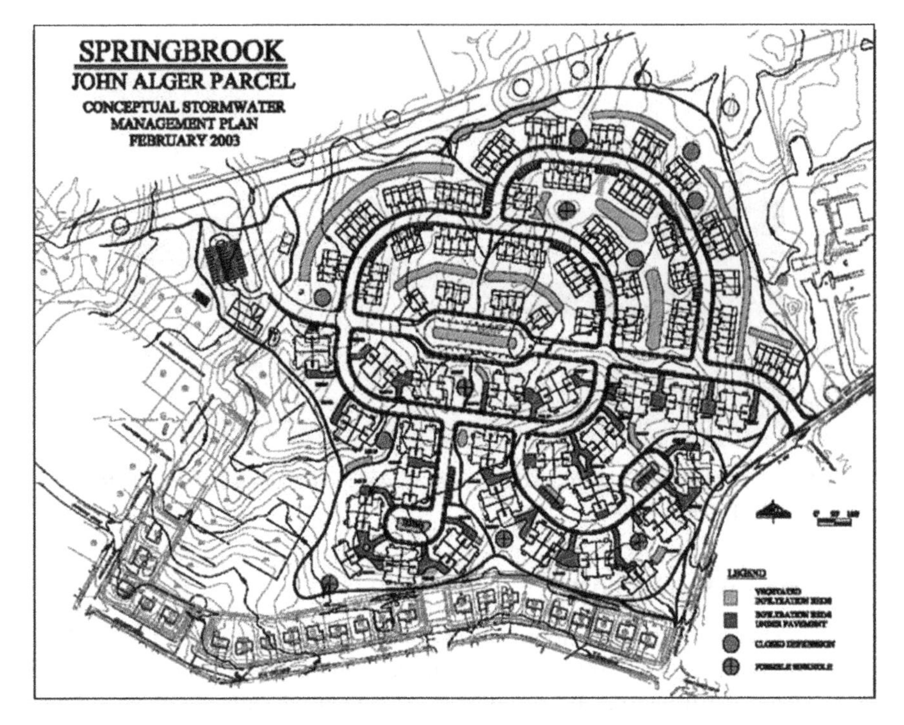

Figure B-36 Porous AC pathways and porous PCC sidewalks were used.

on SUVs failed, and the resulting litigation greatly reduced the company's cash reserves. As a result, the McDonough team was recalled and the project was completed by the Brazilian designers.

University of North Carolina Long-Term Parking Areas (2003–2004)

As part of a campus expansion plan, the University of North Carolina reduced traffic on campus by building large remote parking areas on the outskirts of Chapel Hill. The issue of stormwater management and environmental impacts from construction of large impervious surfaces was critical to both the university and local government. Cahill Associates designed two large parking lots on the north and south sides of town, with both new porous pavement systems and rehabilitated existing impervious pavements, Porous AC and PCC pavements were used to compare material side by side, and all included a stone storage and infiltration bed that infiltrates the soil mantle and recharges the aquifer system, eliminating both increased runoff volume and NPS pollutants (Figure B-40).

The then-existing Estes lot adjacent to the resurfaced impervious lot was expanded by 600 spaces using porous AC for both the parking bays and aisles. Runoff from the remaining impervious surfaces was mitigated by increasing the runoff capture volume from the new beds so that runoff drains through the new pervious AC. At the Friday Center lot on the south side, an additional 1,200

Figure B-37 Infiltration beds use pervious pavements and vegetation.

Figure B-38 Rain gardens were installed on single-family lots.

spaces were constructed with both AC and PCC pavements. This was the first demonstration of this material and method of stormwater management in North Carolina.

West Hollywood Municipal Parking Lot (2004)

In the spring of 2004, the city of West Hollywood, California, unveiled the first porous pavement parking lot in southern California. This small, 28-space parking lot was built on the site of a demolished building on Santa Monica Banlevard, where parking is at a critical level, and was considered by the city to uphold one of its core values of environmental stewardship. The opening was a public relations event, with all media present for the ribbon cutting (Figure B-41).

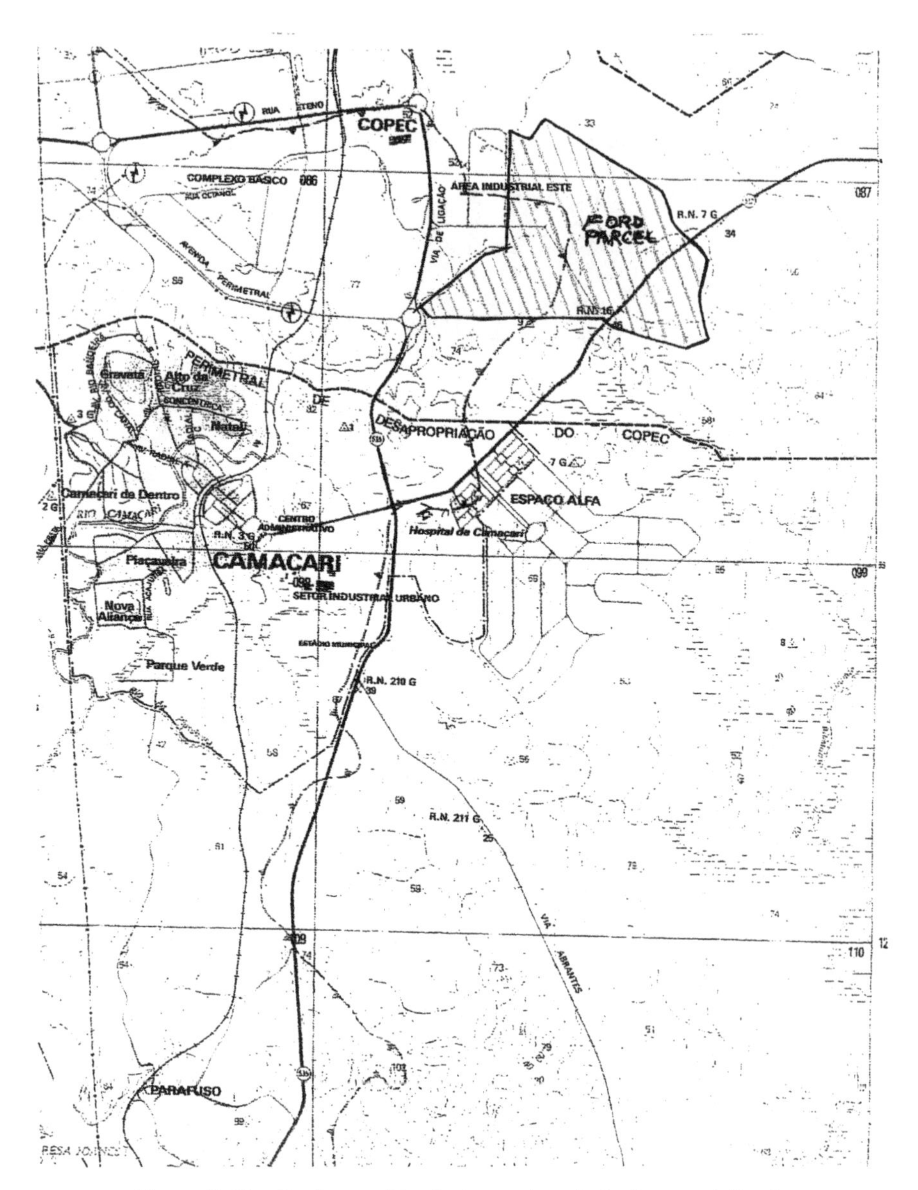

Figure B-39 Ford assembly plant complex site in Camacari, Brazil.

Woodlawn Public Library, Wilmington, DE (2005)

To advance their combined sewer overflow reduction program the city of Wilmington Department of Public Works asked Cahill Associates to incorporate innovative stormwater management practices into the new Woodlawn Library in northwestern Wilmington. In coordination with the city, New Castle County,

Figure B-40 University of North Carolina long-term parking lots used both porous AC and PCC.

Figure B-41 The West Hollywood lot provided a high-visibility site.

and the Friends of the Woodlawn Library, Cahill Associates and Rodney Robinson Landscape Architects transformed the traditional site plan with direct connections to the combined sewer system into a powerful demonstration site for green infrastructure in an urban setting (Figures B-42 and B-43). The project site was the former home of the Department of Motor Vehicles.

Figure B-42 Planter beds infiltrate rooftop runoff.

Figure B-43 Large gain garden provided with signs for children.

Stormwater best management practices (BMPs) include cisterns shaped as children's play blocks, a stormwater planter box, a central bioretention area/rain garden, and a porous asphalt parking lot with storage and infiltration beds. Each BMP is highlighted with illustrative signage that describes the green infrastructure element in greater detail. The project not only reduces Wilmington's largest CSO but educates library patrons, neighborhood residents, the development community, and governmental officials on the importance of stormwater management and ways to incorporate BMPs into sites ranging from residential to urban.

Auto Warehousing Corporation/Port of Portland Terminal 6 Porous Pavement Parking Expansion (2005)

The Auto Warehousing Corporation leases facilities from the Port of Portland to offload and process cars arriving from overseas ports. Due to increases in volume, the facility at Terminal 6 was in need of expansion to resurface some 50 acres for additional auto storage (Figure B-44). Cahill Associates, along with Century West Engineering and GreenWorks, designed a sustainable site and stormwater strategy utilizing porous pavement, infiltration beds, and vegetated swales. In the end, the project entailed some 36 acres of porous pavement (Figures B-45 and B-46), and the use of on-site infiltration facilities allowed for a significantly shortened design and permitting process that made possible construction during the 2006 construction season by avoiding the need for U.S. Army Corps of Engineers or Oregon Division of State Lands permits for an outfall to the Columbia River. The use of porous pavement also allowed the port to avoid the city of Portland's stormwater fees, which are calculated based on a site's impervious area. In addition to the incredible cost and time savings, the porous pavement, infiltration beds, and vegetated swales continue to improve the overall water quality in the vicinity of the site by mimicking the site's predevelopment hydrologic cycle.

Figure B-44 Imported autos are offloaded from ships and stored temporarily.

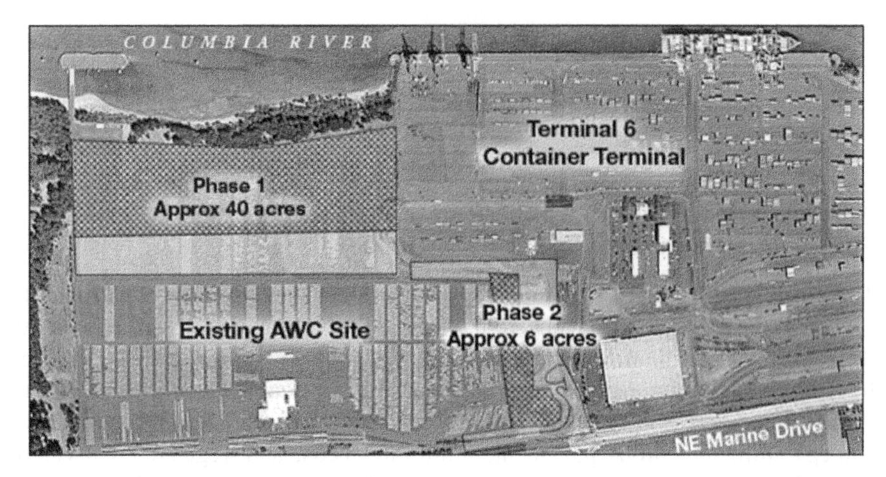

Figure B-45 The Port of Portland site is the largest porous pavement in the United States.

Figure B-46 The size of the lot can be appreciated from this perspective.

San Diego County Stormwater Management Demonstration Project, County Center Complex, San Diego, CA (2006)

San Diego County, California, was interested in developing new solutions for stormwater management that would demonstrate compliance with evolving state and regional criteria for both volume reduction and water quality, as contained in permit requirements. A portion of the existing site was reconstructed with porous AC, porous concrete and paver blocks (Figure B-47), and the contiguous pavements left in their original form (rather deteriorated). A monitoring system was

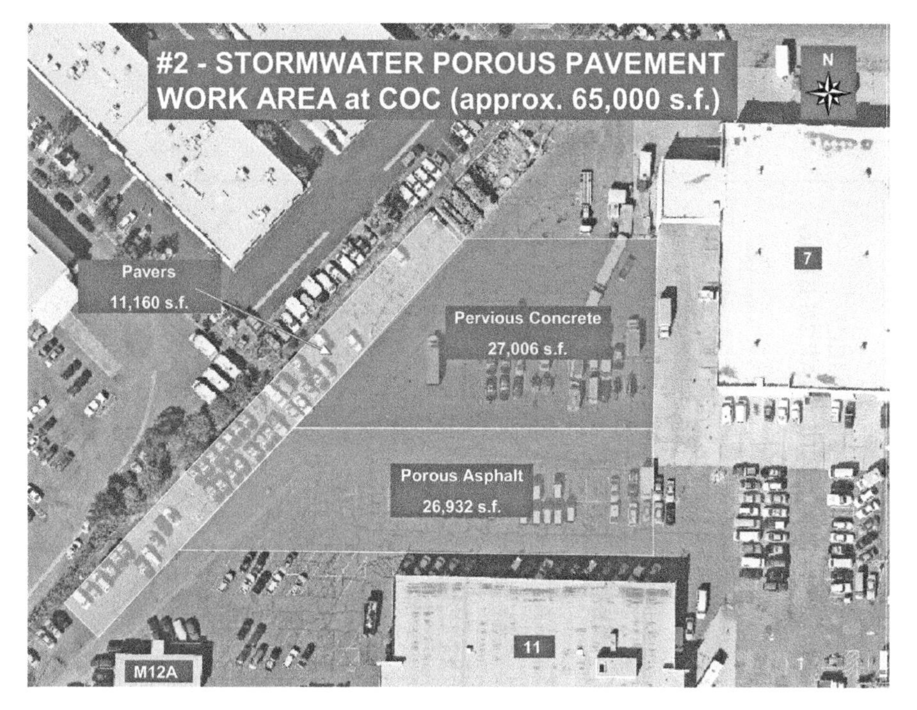

Figure B-47 Different porous pavements were constructed side by side.

designed and installed for both the new and existing pavements, including flow measurements to evaluate water balance and any overflow discharge. In addition, the county installed several subsurface water quality stormwater treatment units downgradient from this area, and again their performance was monitored. All three pavement types were successful in allowing infiltration of rainfall, and the major differential consisted of the cost and maintenance requirements. The results helped to inform the county and other regional governments as to the effectiveness of porous pavements in southern California. In a subsequent demonstration phase, roof runoff was monitored and discharged to the subsurface beds built to serve adjacent buildings.

San Diego County Park, Flynn Springs, El Cajon, CA (2006)

In 2006, Cahill Associates was retained by the San Diego County Department of Parks to retrofit existing parking areas at the Flynn Springs site, as a demonstration of the technology of porous pavement for groundwater recharge and runoff mitigation. In a very arid region where annual rainfall is less than 10 in. a year, the importance of aquifer recharge cannot be overestimated. The county received a grant from the California Water Resources Control Board to complete the retrofit, which not only included the design and installation of the porous

FLINN SPRINGS COUNTY PARK POROUS PAVING PROJECT: POROUS PARKING LOTS WITH GROUNDWATER RECHARGE BEDS

Benefits of Porous Pavements with Groundwater Recharge Bed

- Reduces Stormwater Runoff & Erosion
- Reduces pollutants washed from the land surface by stormwater runoff
- Recharges the Groundwater Aquifer
- Protects the streams from drying up
- Reduces Streambank Erosion and Sedimentation
- Maintains Stream Water Quality and Health

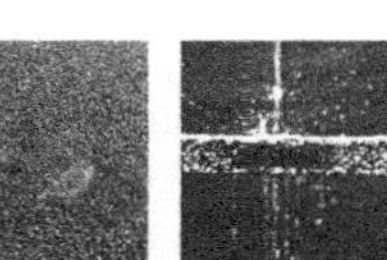

Porous Asphalt
See porous asphalt lot on location map

Porous Concrete
See porous concrete lot on location map

Porous Parking Lot Section with Underground Components

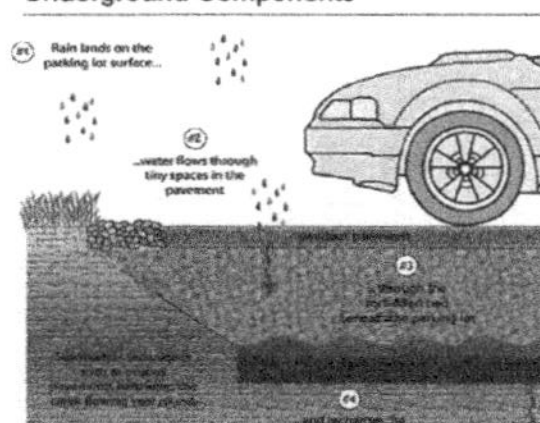

Location Map

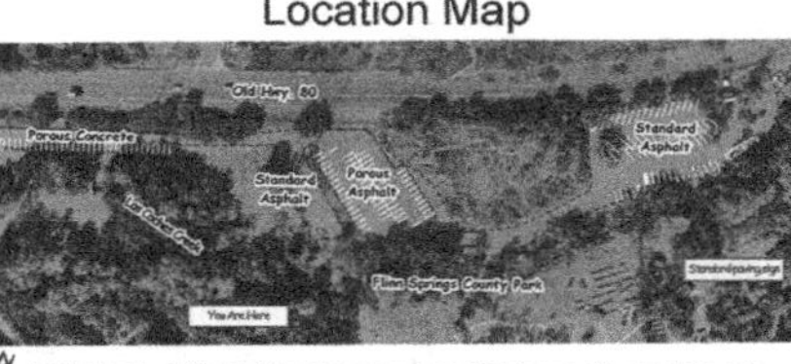

Look closely at the parking lot to your right. This is one of two parking lots that are made with porous pavement. During a rainstorm, water will soak through the surface of the parking lot, and will ultimately recharge the groundwater.

What is different about this parking lot?

Some of the parking lots here at Flinn Springs Park have been paved with *porous asphalt and porous concrete*, which allow rainwater to soak through the pavement and into a storage ground water recharge bed below the surface. Rather than letting rainwater flow off the surface, these parking lots act like a sponge and absorb stormwater, allowing it to soak into the groundwater aquifer – as it did before we constructed standard impervious parking lots and buildings.

Because the pavement is porous, it can't be used everywhere (gas stations, loading docks, etc.) and requires different maintenance such as mechanical street sweeping and vacuuming to keep the pores open.

Maintaining groundwater recharge through infiltration of stormwater is critical to keep Los Coches Creek flowing.

Construction Sequences of Porous Paving Porous Asphalt Lot - View to South-East

Construction Step 1: excavation of sub - grade creating flat terraces on sloped lot for stone recharge bed.

Construction Step 2: Placing geotextile fabric and stone on terraced sub - grade for groundwater recharge bed.

Construction Step 3: Placing porous asphalt over stone recharge bed.

Construction Step 4: Rolling and allowing asphalt to harden before stall line marking, placement of wheelstops, and opening for use by the public.

Figure B-48 The Flynn Springs site included water quality sampling.

pavement beds but also the design and installation of a runoff monitoring system to assess the performance of the system in terms of quantity and quality. The project included 38,000 square feet of porous AC and 12,000 square feet of porous concrete, designed to capture and treat runoff from the existing pavements and some offsite runoff as well (Old Highway 80). The design included vortex chambers to supplement pretreatment of the highway runoff prior to discharge to the infiltration beds.

The runoff monitoring program was developed in conjunction with Kinnetic Laboratories, Inc. The monitoring program demonstrated that there was no increase in runoff volume, and no pollution discharge to the adjacent stream, following renovation. Interpretive signage was used to show the user public how the system worked and the net benefit to water quality (Figure B-48).

Rams Head Project, University of North Carolina (2003)

Cahill Associates, Andropogon Associates, and Roofmeadows were retained as part of the design team for this project. The University of North Carolina had a number of programmatic needs on campus, including additional structural parking, gym space, food services, and clothing stores. They elected to combine all the uses and functions in one new building situated between classrooms and the dormitories south of the main campus, across a wide ravine. The Rams Head building was a major project that included a 600-car parking garage, with a "plaza" on the roof, serving as a pedestrian pathway. The roof was designed as a vegetated roof system, with stormwater capture from upper rooftops and storage in beds and cisterns beneath the lawn open space (Figures B-49 and B-50).

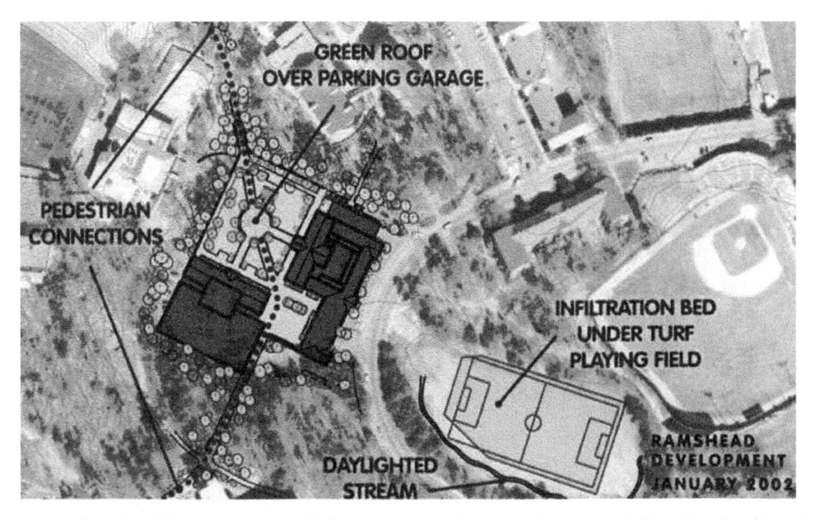

Figure B-49 Rain from upper roofs feeds the roof vegetation, with infiltration in a playfield bed downstream.

Figure B-50 Vegetated roof used storage beneath the surface.

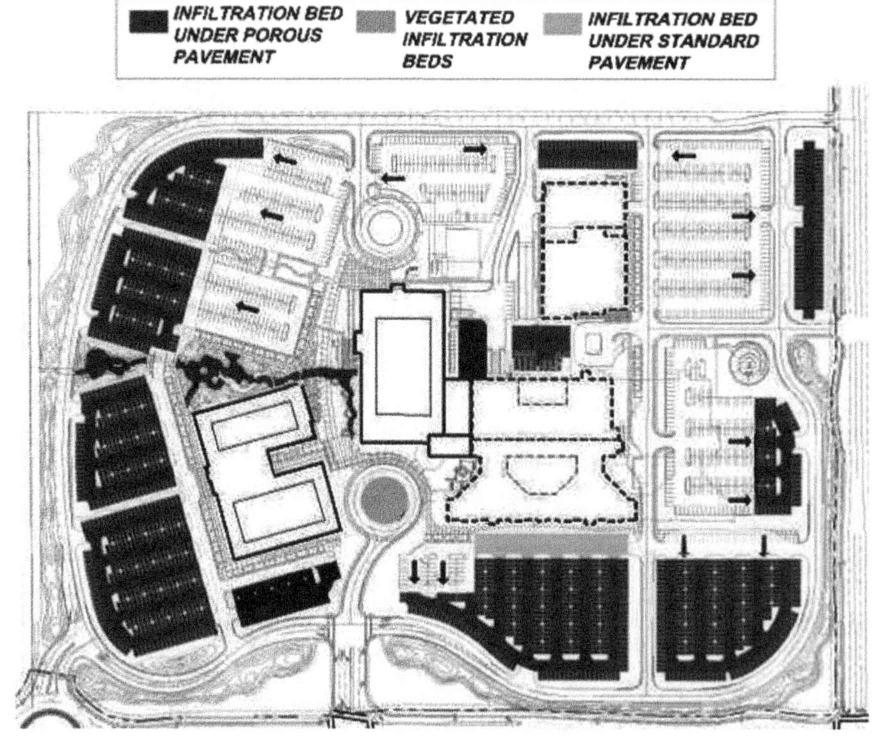

Figure B-51 A 50-acre hospital site used 10 acres of infiltration beds, resulting in zero runoff during the 100-year rainfall.

Figure B-52 Porous pavement testing with a water truck.

The building was constructed on the site of an existing surface parking lot, with a small stream tributary flowing beneath the structure and partially daylighted. The hillside woodland vegetation was protected, and directly downstream a play field was reconstructed as an infiltration system with an open stream channel constructed around it. The historic springs, previously piped, were conveyed in this new channel. Overall, the design solved major university needs in a relatively small and constricted portion of the campus.

Kaiser Permanente Hospital Complex, Modesto, CA (2003)

The town of Modesto, situated in California's San Joaquin Valley, is in the heart of the greatest agricultural area in the country. With a year-round mild climate, an average rainfall of only 12 in., and some of the world's richest soil, Modesto considers itself a city of "water, wealth, contentment, and health." Kaiser Permanente, a 60-year-old public health care provider, constructed a regional hospital and wished to comply with both LEED and evolving California criteria for stormwater management. A stormwater management system was designed using both porous pavement/groundwater recharge beds and vegetated recharge beds (Figures B-51 and B-52). The 100-year frequency rainfall can be infiltrated in the 10 beds constructed, a critical factor in a hydrologic region that has experienced continuous reductions in groundwater levels, due to agricultural demands, and where the depletion of the aquifer system is a major water resource issue.

INDEX

Low Impact Development and Sustainable Stormwater Management, First Edition. Thomas H. Cahill.

Printed and bound by CPI Group (UK) Ltd, Croydon, CR0 4YY

23/04/2025

14660907-0004